Transactional

SIX
SIGMA
and LEAN
SERVICING™

Leveraging
Manufacturing Concepts
to Achieve
World-Class Service

Transactional
SIX SIGMA and LEAN SERVICING™

Leveraging Manufacturing Concepts to Achieve World-Class Service

BETSI HARRIS EHRLICH

S^t_L

ST. LUCIE PRESS

A CRC Press Company
Boca Raton London New York Washington, D.C.

Library of Congress Cataloging-in-Publication Data

Ehrlich, Betsi Harris
    Transactional Six Sigma and Lean Servicing : leveraging manufacturing concepts to
achieve world class service / Betsi Harris Ehrlich.
         p. cm.
    Includes bibliographical references and index.
    ISBN 1-57444-325-9
    1. Service industries—Management. 2. Total quality management. I. Title.

HD9980.5 .E35 2002
658.4′013—dc21
                            2002018976

Visit the CRC Press Web site at www.crcpress.com

Dedication

To my loving parents, Paul and Dixie Plummer

Preface

> During the early part of World War II, the statistical methods used by statisticians for years were successfully applied to the control of quality in production. [These methods were] pioneered by Dr. Shewhart of the Bell Telephone Laboratories. ... About 10 years ago the technique was successfully applied to office operations. ... To many individuals, the comparison between production operations and office operations is difficult. It is easy to understand the sampling of parts off the production line to determine the number of defective parts. However, when converting units of production and number of defective parts to office applications, the analogy becomes somewhat hazy. It has been argued that the two are not the same.
>
> *Systems and Procedures: A Handbook for Business and Industry,* 1959

The more things change, the more they stay the same. Nowhere is that truer than in quality management. When I started in this field more than 20 years ago, Quality Circles — under the umbrella of "participative management" — was the latest management rage, courtesy of Dr. Joseph M. Juran. Working as an industrial engineer at a large bank, I received the mandate from our division head, who most likely read the latest edition of *Fortune* touting this breakthrough strategy of the world's top corporations. Consultants were hired, plans made, and expectations raised.

This cycle repeated itself, in one form or another, for the 20-plus years since and continues to this day. The latest breakthrough, of course, is Six Sigma.

The average life of a breakthrough strategy along the lines of Quality Circles, Total Quality Management (TQM), and Reengineering is approximately 10 years after popularization, with interest peaking at about the 7th or 8th year. Bill Smith at Motorola started the Six Sigma effort in the 1980s; however, it was not until the early 1990s that Larry Bossidy at Allied Signal and later Jack Welch at General Electric (GE) popularized it as a breakthrough business strategy. This being 2002, I debated whether I should write a book on an "outdated" concept when perhaps I should be researching the *next* breakthrough strategy revolutionizing the world's top corporations.

My answer is yes, no, and maybe, interpreted statistically as maybe. Yes, Six Sigma will no doubt be among the postmortems of the next generation. No, I am not wasting effort in being among those who document and add to the Six Sigma knowledge bank. And finally, as a professional in the field, perhaps I should be researching the breakthrough strategy that will take American companies to the next level of global competitiveness.

THE CONTINUOUS JOURNEY

In the spirit of the profession, the study of quality management is a continuous journey of improvement, much like the processes we continually strive to improve. Who can argue that each new breakthrough strategy does not improve upon the previous one?

Our journey begins with the ancients debating whether the planets revolved around the sun or the other way around. Tycho Brahe had the unusual idea to solve the debate by taking measurements that were accurate enough to determine the motions of each.* Previous debates centered on philosophical arguments. Brahe's radical idea to study a problem by measurement and experimentation, rather than opinion, began the journey toward the scientific method.

The journey continues through the age of enlightenment with the likes of Pascal, Gauss, and Poisson, among others, developing and refining probability and error theory. Galton's studies of hereditary traits in the late 19th century led to the development of correlation and regression, establishing the basis of modern statistical methods that were further developed by Pearson, Fisher, Gosset, and Spearman.

From modern statistics came the industrial quality control techniques of Shewhart, Dodge, and Romig. After World War II, when American industry shunned him, Deming took his statistical methods of continuous improvement to Japan, giving them the tools to easily surpass the United States in the quality wars. Juran, Ohno, and Taguchi were significant contributors to the quality movement during the post-WWII years, developing many of the basic quality tools used today.

Once Americans realized the inequities in manufacturing quality with Japan, we quickly turned around our factories by applying Japanese quality techniques, American style, enabling the United States to achieve one of the highest standards of living in the world.

From Deming and Juran came the idea that those closest to the process are in the best position to improve it. Quality Circles was the first widespread adaptation of this concept, followed by TQM, Continuous Process Improvement (CPI), and Business Process Reengineering (BPR). Each new strategy was a variant of the previous, recognizing their weaknesses and striving to improve.

Six Sigma builds upon many of the successful elements of past breakthrough strategies and incorporates unique methods of its own. Most prominent of these are the rigorous quality targets (six sigma equals 3.4 defects per million opportunities) and an emphasis on measurements and statistical methods.

Just as our profession urges a "better, faster, and cheaper" product — what I refer to as the velocity of improvement — Six Sigma will no doubt yield to a better, faster, cheaper breakthrough strategy. If it does not, our future generation of quality management professionals will have failed.

That said, I firmly believe that all significant quality improvement strategies for the past 40 years are variants of Deming's original Plan–Do–Check–Act continuous improvement strategy. They all use the scientific method to find and fix problems on a continual basis, usually within a team environment. The tools and techniques may have become more sophisticated along the way, but the basic approach remains the same.

---

* Feynman, Richard P., *Six Easy Pieces: Essentials of Physics Explained by Its Most Brilliant Teacher*, Persus Books, Reading, MA, 1995, p. 90.

TRANSACTIONAL SIX SIGMA AND LEAN SERVICING™

Service industries have traditionally lagged manufacturing in adoption of quality management strategies, and Six Sigma is no exception. Its genesis was when Motorola engineer Bill Smith recognized that field failures of a product were correlated to the amount of rework needed during the manufacturing process. From this came the goal of driving out nonconformities through achieving a six sigma level of defects. It was not until a decade later that Six Sigma was widely applied to service processes at GE.

Having the unique advantage of applying practically all of the latest strategies in servicing environments over the last 20 years, in addition to keeping up with leading-edge manufacturing quality methods, I am impressed with what I believe is the next generation of breakthrough strategy: Lean Manufacturing, and more specifically Lean Servicing™, which is applying lean concepts to transactional and service processes.

Lean Manufacturing evolved from the Toyota Production System (TPS), developed during the post-WWII years, incorporating concepts borrowed from Henry Ford's Highland Park and River Rouge Model T factories. Ford named his mass production system "flow manufacturing," a central concept of Lean Manufacturing. Ford's simple key concepts of a conveyor line to move work to the worker, specialization of labor to separate work into simple repetitive tasks, and an integrated supply chain to bring materials to the assembly line share many traits with lean. Ford's mistake was to arrange his fabrication machinery into separate workstations, which translated into batch production, queues, inventory, and ultimately a "push" scheduling system.

Taiichi Ohno of the Toyota Motor Company blended Ford's concepts of flow manufacturing with his own unique ideas, such as elimination of all forms of waste, including time, materials, and space; *jidoka* (built-in quality); *gemba* (action on the production line); and "pull" scheduling, meaning that the product is built to actual demand rather than forecasts. Continually improved since inception, TPS is synonymous with Lean Production and has made Toyota one of the most profitable companies in the world.

I have applied the concepts of Lean Production in transactional environments to achieve significant improvements in a system I have coined Lean Servicing™. The concepts of built-in quality, continuous flow, and small-batch production, along with eliminating all forms of waste are especially adaptable to transactional processes, such as an insurance claim moving through the authorization process.

INTENT AND STRUCTURE OF THE BOOK

This book is intended for those new to Six Sigma and Lean Production concepts, as well as the more experienced quality professional. Its primary purpose is to serve as a practical guide and reference for practitioners implementing Six Sigma in a transactional environment.

Case studies are included to help the reader understand how Six Sigma and Lean Servicing™ are successfully applied in familiar situations. Liberal use of examples, graphics, and tables also assist in explaining these often difficult concepts.

Specially dedicated "technical zones" can be skipped with no interruption in the main text. These may be of interest to the more technical and experienced reader as well as serve as a reference for all readers when applying a particular tool or technique.

Chapters 1 through 5 describe the foundation for successful implementation of any broad company initiative, covering project and change management in addition to working within teams. A history of quality management is provided to assist the reader in gaining a deeper understanding of the long journey that has preceded where we are today.

Chapters 6 through 10 describe the Define, Measure, Analyze, Improve, and Control (DMAIC) process, the cornerstone of Six Sigma. Design for Six Sigma (DFSS) is covered in Chapter 11. The tools and techniques are described, along with examples and lessons learned from the trenches.

Chapter 12 is an introduction to Lean Servicing™ and how it can be utilized to achieve the next level of service excellence and ensure increasing velocity in our never-ending journey of continuous improvement.

Acknowledgments

Many individuals have assisted in creating this book. Most of all I want to acknowledge my family: my husband Carl and sons Adam and Andrew. Their assistance, understanding, and patience are greatly appreciated. Much appreciation goes to Rebecca Fay for her assistance and inspiration in seeing this through. I wish to acknowledge Tricia Nichols and Carrie Roediger for keeping my spirits up when I needed it most. I would like to acknowledge the staff and volunteers of the American Society of Quality, especially Jim McClinn and Ronald Atkins, for providing the opportunity to be involved with the Six Sigma Black Belt certification process, leading to a greater appreciation of the Six Sigma improvement approach. A special word of appreciation goes to Forrest Breyfolge, III, and his staff at Smarter Solutions, Inc., who provided my Black Belt training. Forrest's guidance and enthusiasm for Six Sigma opened my eyes to its many possibilities. And finally, thanks to Drew Gierman, Erika Dery, and Amy Rodriguez at CRC/St. Lucie Press for their assistance and patience throughout the publishing process.

Betsi Harris Ehrlich

Author

Betsi Harris Ehrlich has more than 20 years experience as a professional industrial engineer, primarily in financial services and health care. She graduated with highest honors from the Georgia Institute of Technology with an industrial engineering degree concentrating in the healthcare industry and holds a Masters of Business Administration with a specialty in finance from the University of Connecticut. Ehrlich has spent the majority of her professional career as a technical specialist and leader in quality and process improvement and has gained a unique perspective on the evolution of the quality management field from first-hand experience. In addition to being a practitioner, Ehrlich has experience as an adjunct professor, developing curriculum and teaching graduate and undergraduate classes in statistical process control and quality management. Ehrlich is certified as a Six Sigma Black Belt by the American Society for Quality (ASQ) and is recognized by ASQ as a subject matter expert in the field of Six Sigma, having participated in the development and piloting of the Black Belt certification exam offered by the society. She has authored numerous technical publications and newspaper columns on a variety of topics, including quality management. Ehrlich is listed in Marquis's *Who's Who in America* and *Who's Who in the World*, the American Biographical Institute's *2000 Notable American Women*, and the International Biographical Centre's *World Who's Who of Women* and has received many honors relating to her professional career.

Introduction

Time keeps on slipping, slipping, slipping into the future. ... I want to fly like an eagle, to the sea, fly like an eagle ... let my spirit carry me. ... I want to fly right into the future. ...

"Fly Like an Eagle," The Steve Miller Band, 1976

The corporate model we have spent the past century developing and improving will almost surely disappear with this generation. Vast changes in socioeconomic and political forces guarantee the demise of the modern corporation — the monolithic structures we have all despised at some point, usually after spending too much time and mental energy desperately trying to solve a rather simple issue with an under-trained, underpaid, and misrepresented "customer service representative."

Employment growth for the past 20 years has been in mid- and small-size companies, with the big corporations' growth rates steadily decreasing. Although it has always been the case that small upstarts displace entrenched industry leaders (the average life of a corporation is only 30 years),[1] the trend has accelerated in recent years. In addition to the accelerating shift in power between upstarts and veterans, the methods by which market leadership is attained are very different than in past years. Industrial-era management from the mass-production days will surely be a page in the history books by mid-century. The internet and its proliferation of dot-com successes and failures are the tip of the iceberg compared to the unorthodox business climate that lies ahead.[2]

Innovation, flexibility, and rapid change — are they the keywords or the buzzwords for 21st century business? Probably a little of both. Companies likely to succeed are those that transcend the latest breakthrough management strategy. Looking beyond marketing campaign copy, these companies share traits of innovation and flexibility (buzzwords) as well as strong leadership, solid operational perfor-mance, and plain old-fashioned luck (keywords).[3]

This book is about solid operational performance — specifically, how to achieve it. It involves a multidisciplinary approach, combining Six Sigma quality and lean production methods to achieve lasting profitability.

Japanese carmakers, specifically the Toyota Motor Company, long ago realized the simple fact that *product quality is the result of process quality*. It is astonishing that after years of evidence to this fact, most companies still do not understand this concept. Although many are quick to tout their latest breakthrough management strategy to produce high-quality products at a competitive price while empowering their employees to offer their stakeholders the value supply chain through gap analysis of vertical markets, and so forth and so on, translating theory into action and subsequent results is difficult at best.

One of the first American CEOs to understand the importance of process in producing high quality was Harvey Golub of American Express (Amex). Golub is credited with saving the company from the brink of disaster in the early 1990s, when VISA was rapidly gaining market share and arrogance alienated their most important partner — the service establishments who accepted the card. Golub, a former McKinsey consultant, understood that *process management* was the key to saving the company. The elitist strategy of his predecessor, Jim Robinson, had outlived its time; credit cards had become a commodity, and the low-cost, high-quality provider was winning.

Golub had the vision to understand that only through competing head-on with VISA, which meant competing on *costs* as well as *service*, could he save the company. In 1992, he declared war on VISA and stated that Amex would take a billion dollars out of their costs in the next 3 years. He achieved this goal not through sexy marketing or hostile takeovers and mergers, which had captured the attention of the corporate titans of the prior decade, but rather through superior process management. Golub also successfully guided Amex through the industry's movement toward commoditization and returned the company to double-digit profitability. Today American Express is accepted at gas stations, WalMart, and K-Mart, something no one would have thought possible only a decade earlier.

Another well-known company adopted superior process management as their competitive strategy, leading them to become the most successful company in the world in terms of wealth creation. General Electric (GE), under the leadership of Jack Welch, embraced Six Sigma as the way to implement the strategy.

TRANSACTIONAL SIX SIGMA AND LEAN SERVICING™

Adopting any kind of company-wide initiative requires a great deal of hard work, and Six Sigma is no different. Most companies that have been around for more than a couple of decades have probably experienced a company-wide improvement initiative along the lines of Zero Defects, Total Quality Management (TQM), Reengineering, and Six Sigma. Many have been quite successful, such as Amex and GE. Others have been less so, such as DEC and Westinghouse.

Postmortem studies have attempted to diagnose the factors that lead to either success or failure. I suspect that it is not as black and white as the business media would have us believe but, like most things in life, many shades of gray. To start with, the degree of success and failure depends on how one measures success and failure, or what Deming coined the "operational definition." This concept is further explored in Chapter 7.

Suffice to say, there are degrees of success in implementing company-wide initiatives and there are certainly common themes around those degrees. Chapters 1 through 5 deal with these common themes. Analogous to the process of building a house, building the framework of any broad initiative is an absolute must — teaching the quantitative tools of Six Sigma on day 1 does not make any more sense than bringing in the electricians before the land is cleared and the foundation laid.

A NOTE TO READERS

One of the many advantages Six Sigma has over its predecessors is the practice of training business line employees in the quantitative skills required for analyzing and improving complex problems. By line employees, I mean those who have managerial responsibility for the financial results of a particular business unit, as opposed to staff employees, who support the company through specialized knowledge in a particular field such as finance or engineering.

This practice, however, is an advantage only if the line folks are appropriately trained and can apply the tools to achieve practical results. Providing sophisticated statistical tools to someone with no background in statistics is unlikely to result in bottom-line benefit; the skill set has to be matched to the job at hand.

For this reason, this book offers widely used tools that are less difficult to understand (not necessarily easy) and therefore have a greater probability of successful application. For the more experienced engineers and statisticians who are in a Six Sigma role, the book includes designated "technical zones," which offer more advanced statistical methods. It has been my experience that the mainstream tools are sufficient for about 80% of situations, whereas the more advanced tools are needed in the remaining 20%.

If you do not have the interest or the time to learn and comprehend the more advanced statistical tools in the technical zones, I advise you to consult with an expert when these are needed. Most companies who undertake Six Sigma have Master Black Belts who are highly trained in statistics. Many web sites have bulletin boards with participation by highly trained Six Sigma experts who are more than willing to share their expertise and assist with the more complex issues.

A NOTE ON STATISTICS

In my 20 years of teaching statistics and applying statistical techniques in a business environment, I have frequently encountered an almost irrational fear of statistics, often from highly skilled experts in professions such as medicine or finance. I have concluded that some of this fear stems from the terms and symbols used, referred to as statistical notation. Liberal use of an outdated lexicon to represent our 21st century business situations is sure to cause confusion among all but career statisticians.[3] Anyone involved in Six Sigma should take the time to learn the basic terminology and concepts of statistics. The goal of writing this book is to make statistical and measurement concepts readily accessible to those who need it the most — the employees who are the first line of defense for winning customers for life.

REFERENCES

1. Collins, J.C., *Built to Last: Successful Habits of Visionary Companies*, Harper Business, New York, 1997.
2. Drucker, P.F., *Management Challenges for the 21st Century*, HarperCollins, New York, 1999.

3. An example is the term "coefficient." To the layman, coefficient sounds like it means "being efficient together." To the statistician, coefficient is a constant factor of a term as distinguished from a variable.

Table of Contents

1 Introduction

1.1 OVERVIEW OF SIX SIGMA

1.1.1 SIX SIGMA DEFINED

Technically, Six Sigma is a measure of variation that represents 3.4 defects out of one million opportunities for defects. However, it has become much more than its statistical definition. For our purposes, Six Sigma is defined as a *disciplined, data-driven approach of continually improving process quality and productivity to result in bottom line profitability.* It does this primarily through reducing the amount of variation in a process, leading to consistent and predictable output. Let us decompose this definition to gain a greater understanding of this powerful approach:

- *Disciplined.* Six Sigma uses a standardized step-by-step process with specific tools for conducting projects. This is called DMAIC, which stands for Define, Measure, Analyze, Improve, and Control.
- *Data driven.* The emphasis is on using data supported by statistical measures and analysis to make decisions leading toward logical improvements. This is in contrast to making decisions based on opinions, or worse, on fear in a hierarchical command-and-control environment.
- *Approach.* Six Sigma is an approach — a systematic, consistent advancement toward achieving near-perfect quality.
- *Continually improving.* Never-ending improvement.
- *Process.* Repeatable, measurable series of tasks that translate inputs into outputs.
- *Quality.* The capability of a process to meet or exceed expectations.
- *Productivity.* The capability of a process to transform inputs into outputs in an effective and efficient manner.
- *To result in bottom line profitability.* Six Sigma is a business-driven initiative, originating with top CEOs such as Bob Galvin of Motorola and Jack Welch of General Electric (GE), who are judged by financial results. This contrasts to past quality management strategies, which were largely driven by the quality department with often fuzzy goals and results.

Other key concepts of Six Sigma include:

- *Customer-centric approach.* Dr. W. Edwards Deming was among the first to understand that customers define quality. Voice of the customer is a

Six Sigma technique used to determine critical-to-quality (CTQ) attributes of a product or service.

- *Process focus.* The Japanese are well known for transforming themselves from low-quality junk producers to high-quality, world-class producers. Today, the label "Made in Japan" conveys a very different image than 40 years ago. Japanese manufacturers made this transformation by understanding that the only way to improve the quality of a product is to improve the quality of the process used to make the product. American manufacturers have been slow to catch on to this concept.

- *Process capability.* The ability of a process to meet customer CTQs is a key concept in Six Sigma. Determining whether a process is capable involves understanding first, the customer requirements, and second, the process performance. Both require a measurement system with a high degree of reliability.

- *Variation.* The only method of achieving a six sigma quality level (i.e., 3.4 defects per million opportunities) is to reduce process variation. The primary tool used to control process variation is the control chart, first developed in 1924 by Walter Shewhart at Bell Labs. GE, one of Six Sigma's biggest proponents, offers, "Our customers feel the variance and not the average."[1]

- *Defects.* Emphasis is placed on preventing defects through robust process design and ongoing control through the use of control charts and mistake-proofing methods.

- *Infrastructure.* Six Sigma involves a well-defined infrastructure. Six Sigma projects use DMAIC — a systematic methodology that subscribes to the scientific management approach, known in the business world as fact-based management. Scientific management involves collecting data to make an informed hypothesis about something and using statistical tests to either prove or disprove the hypothesis.

1.1.2 SIX SIGMA BACKGROUND

Motorola engineer William Smith, known as the father of Six Sigma, coined the term Six Sigma in 1988, although he worked on the concepts for many years prior.[2] Although everyone talked about preventive quality techniques in the 1970s and '80's, few companies actually adhered to it. Instead, supplier acceptance sampling and visual inspection were the norm for ensuring quality. Defects discovered during the quality control process were corrected through rework and shipped on to the customer.

Smith was the first to use data to prove that products built with fewer defects (i.e., less rework) performed better over the life of the product. He presented his findings to Motorola executives, who at the time were bidding for the Malcolm Baldrige National Quality Award (MBNQA). Bob Galvin, Motorola's chairman, was impressed by Smith's work and provided the resources necessary to integrate the Six Sigma quality methodology into Motorola's operations. One of those resources was Mikal Harry, who was tasked with diffusing Six Sigma throughout the company.

Harry received funding to start the Six Sigma Research Institute at Motorola's Chicago headquarters.[3]

Harry realized that success in spreading Six Sigma among 52 worldwide locations with more than 100,000 employees depended on rapid knowledge transfer, and he concluded this could not occur within the quality department. Instead, specially selected, high-potential employees from throughout the organization were provided intensive training in the Six Sigma methodology.

Employees were given martial arts titles. Those who were selected for full-time Six Sigma roles were called Black Belts, whereas those who stayed primarily in their jobs while working on Six Sigma projects were called Green Belts.[4]

Six Sigma caught on early at other large manufacturers, such as Asea Brown Boveri (ABB) and Allied Signal, now owned by GE. The success of Six Sigma in raising quality levels while simultaneously lowering expenses caught the eye of GE Chairman Jack Welch. GE began implementing the approach at their manufacturing sites in 1995 and later at their financial services divisions. GE consistently cites Six Sigma as the force behind their record profits and ability to become the largest market-capitalized company in the world.[5]

1.1.3 TRANSACTIONAL VS. MANUFACTURING SIX SIGMA

All significant quality management initiatives have originated in the factory. From Shewhart's quality control charts to Motorola's Six Sigma initiative, the emphasis has been on the quality of the product. Even when the Japanese, under the guidance of Deming and Juran, began to emphasize the importance of process quality, it was directed at manufactured products.

Early efforts at service quality management and improvement were categorized as work simplification initiatives in an office environment.[6] In the late 1940s, United Airlines reduced their reservations error rate from 30 to 5% in a 6-week period through a combination of work simplification and quality control charts. In 1950, the Standard Register Company applied similar techniques to their billing process, significantly raising accuracy rates with one half the previous efforts. Prudential Insurance Company's successful application of Statistical Quality Control (SQC) was described in a 1953 article in *The Office Executive* magazine titled "Prudential's Program for Clerical Quality Control."[7]

1.1.4 TRANSACTIONAL SERVICE QUALITY

A distinguishing feature of transactional service delivery is the high volume of employee–customer encounters, known as "moments of truth," which are absent in a factory or back-office setting. This brief human-to-human contact requires a different perspective of quality. Traditional measures, such as percent defective and fitness for use, are not appropriate in a transactional environment. The emphasis on precise, statistics-based measurement of defects serves manufacturing well, where production of a tangible product encourages a high degree of scrutiny of documented specifications. Moreover, back-office service process measures, such as cycle time and accuracy, cannot fully capture the true measure of whether the customer's CTQ requirements are satisfactorily met.

Additionally, a tangible product or information printed on a billing statement can be reworked, with the end result a defect-free product, if not a defect-free process. This is not the case with transactional service quality. The time spent on a phone call, and to a large extent, the customer's satisfaction with the call, is forever lost. Although service recoveries are possible, they are expensive and usually insufficient to overcome the initial negative encounter. Responsiveness, knowledge, ability to help, and making the customer feel valued are the new measures of quality in a direct customer transactional environment.

The process of manufacturing transforms raw materials (input) into product (output); the quality of the incoming materials significantly contributes to the quality of the finished product. For this reason, acceptance sampling of incoming lots has been a widespread practice in manufacturing. Fortunately, more companies are developing supplier partnerships in which incoming quality of materials is guaranteed through certifying the quality of the supplier processes in making the materials.

What is the input into a typical transaction service delivery? An employee's ability (input) to transform a customer's problem into an acceptable solution (output) is one. Supporting the employee's ability are processes, and *the quality of these processes determines the quality of solutions.* Among these processes are the recruiting and hiring process, the training process, and the process of delivering to the employee the information necessary to solve the customer's problem. Underlying these processes are the policies that provide the rules the employee follows in solving the customer problem. Ideally, the rules permit a solution acceptable to both the customer and the organization.

To summarize, the distinguishing feature of transactional service quality is that each customer encounter is valuable, making every episode of poor service expensive. High volumes of direct customer encounters require a highly trained workforce whose primary consideration is efficiently meeting the customer's requirements. Whereas raw material quality is the key measure in the factory, workforce quality is the key measure in a transactional environment.

1.2 OVERVIEW OF LEAN MANUFACTURING

1.2.1 What Is Lean Manufacturing?

Lean Manufacturing, sometimes called Lean Production, is a business philosophy that was originally developed at the Toyota Motor Company, where it was called TPS, for the Toyota Production System. The objective is to eliminate all forms of waste in the production process. The Japanese term for waste is *muda*. Forms of waste include the following:[8]

1. Overproduction — using production equipment and machines faster than necessary without regard to whether the output of the machines is required downstream.
2. Waiting for machines or operators — operators may wait for available machines or machines may wait until an operator is available.

3. Transportation waste — unnecessary movement of parts or people around the production facility.
4. Process waste resulting from inefficient, poorly designed processes — duplication of effort, inspections, and nonvalued activities.
5. Excessive inventory — unnecessary work in process and finished goods beyond what is needed on a normal basis to keep the business running. The "evils of inventory" result in wasted floor space, capital, administrative costs, and storage costs.
6. Wasted motions through operators leaving workstations to fetch required supplies or through continuous reaching, searching, or carrying goods.
7. Waste of rework through producing defects.

When we eliminate all waste, the order production cycle time (time from receipt of order to receipt of payment) is compressed. The result is short cycle and delivery times, higher quality, and lower costs.

Lean's many techniques to deal with waste work in synergy, resulting in a whole that is greater than the sum of the individual techniques; this holistic approach requires substantial change and effort throughout the entire manufacturing process. However, properly implemented, Lean Manufacturing produces high-quality products built to customers' requirements at lower costs than traditional manufacturing.

1.2.2 LEAN TECHNIQUES AND TOOLS

The following is an overview of key lean tools and techniques. More detail on each of these as well as others is provided in Chapter 12.

- *Small-batch or single-piece continuous-flow process.* To reduce cycle time and minimize inventory and rework, small batches of work are produced. Single-piece continuous-flow production, where one product is made at a time, is desirable if feasible. Not only does this decrease inventory and its costs, it provides for early detection of defects in the process and thus less rework at final assembly.
- *Just-in-Time (JIT) inventory system.* For small-batch continuous flow to work, a JIT inventory system is necessary. This means that the manufacturer must work in partnership with their suppliers to schedule frequent deliveries of small lots.
- *Pull scheduling.* Traditional manufacturing builds products to sales forecasts, developed by marketing and finance. This is called push scheduling, since orders are pushed through the system and then pushed upon customers through aggressive sales techniques. Large batches are pushed through the manufacturing line, building queues and work in process inventory at each step. Lean adheres to pull scheduling where orders are built to customer request. Raw materials are received in small batches on one end of the continuous-flow system and the order is pulled through the manufacturing process through final assembly and shipped immedi-

ately to the customer. The cycle time from customer request to invoicing and payment receipt is appreciably shorter than push scheduling.[9]

- *Reliable process through Total Productive Maintenance (TPM).* Because Lean Manufacturing builds to customer request in a very short cycle of production, it is essential that all equipment used to make the product is up and running at all times.[9] Lean addresses this through TPM, making equipment operators responsible for preventive maintenance of their machines and giving them the authority to take corrective action if necessary. This in turn calls for a highly skilled and cooperative workforce. Because the operators that work with the machines day in and day out are responsible for their maintenance, they are more likely to understand how the machines can be improved to make their jobs easier. Without high reliability of equipment through preventive maintenance, Lean Production will likely fail.

- *Capable processes.* Lean Production depends on processes that are capable of meeting requirements without rework — doing it right the first time. The goal of lean is to eliminate all forms of waste; making defective products, with the resulting rework and material wastes, is unacceptable.

Six Sigma is the right approach to build defect-free processes. Integrating the tools and techniques of Six Sigma within the larger framework of Lean Production, one can make and deliver high-quality, low-cost, customer-specified products on a sustained, continuously improving basis. This is true whether the product is an automobile or a transaction at an automatic teller machine.

Other tools of lean, including *kaizen*, visual control, the 5S's, Single-Minute Exchange of Dies (SMED), and mistake-proofing, are discussed in Chapter 12.

1.3 SIX SIGMA COMPARED TO TOTAL QUALITY MANAGEMENT AND BALDRIGE QUALITY APPROACHES

1.3.1 Is Six Sigma New?

Critics claim that Six Sigma has added nothing new to the quality management field, but merely rehashes old tools with a catchy new name. The truth is that, like every approach that has preceded it, Six Sigma borrows what has worked and improves on what has not.

The field of quality management has its roots in Shewhart's work at Bell Labs and Western Electric. Shewhart used statistical methods to understand variation and was the first to distinguish between variation by chance and variation by cause. Applying his methods to reduce variation in the manufacturing process vastly improved product quality.

Since Shewhart, many additional quality tools have been developed, most of them rooted in statistics. From Statistical Process Control (SPC) to Quality Control to Quality Circles to Zero Defects to Total Quality Management (TQM) to Baldrige to ISO9000 to Continuous Process Improvement to Reengineering to Six Sigma to

Lean Production and beyond, all approaches share a common thread first developed by Shewhart: use of data-driven, statistics-based tools to draw conclusions. The approaches to infrastructure and implementation, as well as the particular techniques, have differed, with mixed results. The fact remains that not only do all approaches share common tools, they also share a common goal — improving quality and productivity.

Each successive approach recognizes the weaknesses of the previous one and seeks to improve upon it. This is exactly what we should expect; the profession that is dedicated to improving quality and productivity is simply following its own mantra of continuous improvement.

To this point, those of us involved in the quality field need to focus our energy toward improving today's version of Six Sigma, through understanding what works and what needs retooling, and then taking concrete steps toward making these improvements. Rather than disparaging ourselves about past "failures" we should be proud that we have the introspective skills to see beyond the trees and into the forest.

With that said, some real differences exist between Six Sigma and recent quality management approaches — in particular, TQM and the MBNQA.

1.3.2 TOTAL QUALITY MANAGEMENT

The TQM approach realizes that widespread involvement of those closest to the process is the key to improving the process. TQM has its roots in Japan's Total Quality Control, which originated as Company-Wide Quality Control in the 1950s.[10]

TQM emphasizes teams — both ad hoc improvement teams and self-directed work teams. Following the lead of Dr. W. Edwards Deming, Japan was handily winning the quality wars when TQM emerged in the 1980s. Deming's focus on process and a structured approach toward improvement through the Plan–Do–Check–Act (PDCA) circle of improvement are key elements of TQM.

Where Six Sigma differs from TQM are the specific goals of each and the execution of the strategy. TQM sets vague goals of customer satisfaction and highest quality at the lowest price, whereas Six Sigma sets a specific goal of 3.4 defects per million opportunities. Six Sigma also focuses on bottom line expense reductions with measurable, documented results. Thus, unlike TQM, project selection is of prime importance in Six Sigma.

For execution of the strategy, TQM is owned by the quality department, making it difficult to integrate throughout the business. Six Sigma is a business strategy supported by a quality improvement strategy. Further, the infrastructure of Six Sigma is designed to be owned by the business units and supported by the quality department. Top CEOs such as Bob Galvin of Motorola and Jack Welch of GE have clearly "owned" Six Sigma in their organizations.

Finally, TQM seeks to be all things to all people. At one company, all division heads were instructed to read the same speech at the same time to division employees announcing the start of a TQM initiative. The speech began: "From this day forward, everything is different. ..." These sorts of lofty, fuzzy proclamations, absent of concrete methods and goals, have led to the decline of TQM in American industry. In 2002, it is rare to see an active and viable implementation.

1.3.3 MALCOLM BALDRIGE NATIONAL QUALITY AWARD

The Malcolm Baldrige National Quality Award (MBNQA) is a government-sponsored award recognizing U.S. organizations for their achievements in quality performance. Established by Congress in 1987, its namesake is former Secretary of Commerce Malcolm Baldrige.

The award is given annually to companies in one of the following categories: (1) manufacturing, (2) service, (3) small business, (4) education, and (5) health care. There have been 41 recipients from 1988 to 2001. Motorola was one of three companies to receive the inaugural award in 1988; the newly launched Six Sigma initiative contributed to their success in winning the prize.

Seven categories of quality criteria are used to determine winners:[11]

1. *Leadership.* How do senior executives guide the organization toward quality enhancement and how does the organization demonstrate good citizenship in the community?
2. *Strategic planning.* What is the strategic planning process to determine future direction of the organization?
3. *Customer and market focus.* How does the company determine customer requirements and expectations?
4. *Information and analysis.* How effectively does the company use information and analysis to measure and improve performance?
5. *Human resources.* How does the company effectively encourage its employees to reach their potential and optimize overall company performance?
6. *Process management.* How well does the organization design, manage, and improve its key production and delivery processes?
7. *Business results.* What are the company's performance and improvement in key business areas of customer satisfaction and what is their financial and marketplace performance relative to their competitors?

Both Six Sigma and Baldrige focus on improving value to customers and improving overall company performance. Both also rely on the use of data to guide quality improvement. However, their approach toward quality improvement is different. Baldrige is a descriptive approach. That is, it describes what a quality organization looks like, not how to get there. Six Sigma is a prescriptive approach that describes how to get there, with a great deal of detail, using the highly structured DMAIC approach. There is a subtle but key difference between the two.

1.4 COMMON SIX SIGMA TERMS

A large part of understanding a particular science is understanding the language that is unique to the science. Here are some common terms used in Six Sigma:

- Six Sigma — The statistical measurement of the area outside six standard deviations from the mean value of a normal distribution; this translates into 3.4 defects per million opportunities. Six Sigma also refers to a business strategy or approach for systematic improvement of business processes to achieve bottom line profitability.

- DMAIC — The systematic approach to reducing variation and achieving improvement utilized in Six Sigma. DMAIC is an acronym for Define, Measure, Analyze, Improve, and Control — the five phases of a Six Sigma improvement project.
- Design for Six Sigma (DFSS) — The approach used to design a new process rather than improving an existing process.
- Critical to quality (CTQ) — The characteristic of a process or product that has a direct impact on the customer's perception of quality.
- Black Belts — Specially trained individuals responsible for leading Six Sigma teams through the DMAIC process. They usually occupy full-time positions. Training in statistical methods and quality improvement techniques are a prerequisite.
- Master Black Belts — Individuals who train and mentor Black Belts. They are highly trained in statistics and quantitative analysis, possess excellent leadership skills, and usually occupy full-time positions.
- Green Belts — Similar to Black Belts; however, they may possess less statistical expertise and usually occupy part-time positions.
- Defect – Failure to meet a customer-defined requirement. Defining what constitutes a defect for a particular process or product is the first step in quantifying the number of defects, leading to calculation of a sigma level, which is the number of defects per million opportunities.
- Defects per million opportunities (DPMO) — The number of defects out of one million opportunities for defects.
- Sigma — The Greek letter used to describe the standard deviation of a data distribution. The standard deviation is a measure of the spread of the observations of a data distribution.
- Rolled throughput yield (RTY) — A measure of the amount of rework required in a process, a concept known as the "hidden factory." RTY is determined by multiplying the individual process step yields to arrive at the total process yield before any rework is done.

1.5 A HISTORICAL PERSPECTIVE

This section provides a brief history of probability and statistics. It is interesting to note that 21st century quality methods arose from a dice game popular in the mid-17th century. The odds of rolling a particular set of dice were first calculated by Blaise Pascal and Pierre de Fermat in 1654. This event set into motion the rise of mathematical probability, upon which statistical science is based.

1.5.1 PROBABILITY

Probability involves calculating the chances of a particular outcome, given the total number of outcomes possible. It is commonly expressed as a mathematical ratio; for example, the chance of tossing a head divided by the total number of times the coin is tossed. The dice game that provoked the mathematical calculations of odds by Pascal and Fermat involved the following question:

> *Players A and B each stake $32 on a three-point game. When A has two points and B has one point, the game is interrupted. How should the $64 be fairly distributed between the players?*

Pascal's solution was to divide the problem between mathematical certainty and probability. The certainty was that Player A deserved half of the money, or $32. The probability solution required that Player A receive one half of the remaining $32; thus Player A should receive $48 and Player B $16.

Note that Pascal's emphasis was on *expectation of outcomes* and equality between the two players, rather than calculation of the *probability of outcomes*; a subtle but important difference. Pascal's viewpoint summarizes early probability theory: expected outcomes are based on fairness and value of the outcomes.[12] Fairness is assumed because the conditions of two players are indistinguishable since they are both willing to play. An example of this viewpoint exists today; people buy lottery tickets because of the value of a successful outcome, rather than calculating the risk by computing probabilistic outcomes. The reality is they have very little chance of winning.[13]

The fallacy of this thinking was undiscovered until 1713, when Jakob Bernoulli's work, *The Art of Conjecture*, was posthumously published. Bernoulli correctly viewed probability as a degree of certainty, thus marking a shift away from the *value* of outcomes to the *probabilities* of outcomes. Thus, instead of the dice game being assumed fair, with implicitly equal probabilities, the game is fair because the probabilities of a particular outcome are equal among all players.[12]

Early developers of probability theory sought to predict individual behavior, specifically, the risk an individual was willing to take, based on the assumption of rational behavior. The attempt to mathematically compute the risk-taking behavior of an individual was part of the 17th century enlightenment period, when mathematic certainty was sought for all causes. This was in contrast to an earlier period known as fatalism, during which all of nature was attributed to random acts of God that man could not possibly determine with any certainty.

Although attempts to predict individual behavior based on mathematical principles ultimately failed, this early work led to attempts to quantify the actions of large numbers of people, which was the beginning of the science of statistics.

1.5.2 STATISTICS

Statistics can be viewed as the flip side of probability. Whereas probability seeks to draw conclusions about a particular observation from a larger population (e.g., the odds of drawing an ace of diamonds out of a deck of cards), statistics seeks to infer characteristics of a population based on the characteristics of a sample drawn from the population. The former is based on deductive reasoning whereas the latter uses inductive reasoning. We call the latter inferential statistics.

Unlike inferential statistics, descriptive statistics seek to analyze and quantify large-scale phenomena based on the collective behavior of individual observations.

In 1835, Simeon Poisson noted that individual acts of behavior would cause no significant changes in the larger group; he termed this the "law of large numbers."[14]

Early statistics were purely descriptive. The first documented use of statistics was in 1662 when John Graunt started keeping birth and death records in London. Although individual records had previously been recorded, Graunt was the first to summarize the data in his "Natural and Political Observations on the Bills of Mortality," illustrated in Figure 1.1. Graunt used his records to determine how many able-bodied men in London could serve in the military. The data proved surprisingly consistent in the numbers of births, deaths, crimes, and diseases from year to year.[15]

At first, statistics was called political arithmetic, or "the art of reasoning by figures upon things relating to the government."[16] Almost 200 years passed before statistics evolved from numerical information about society to Poisson's law of large numbers, in which the "state of society" could be predicted based on the uniformity of births, deaths, crimes, and other natural events from year to year.[12]

That the relative frequencies of natural events could be predicted based on the stability of the underlying state, regardless of the unpredictable and irrational actions of an individual, was a milestone in the development of the science of statistics. After Poisson published his law of large numbers, attention turned to quantifying and predicting the statistical behavior of society at large, in which the frequencies of natural events over time conformed to the underlying mean of their probabilities. This period during the first half of the 19th century was when the word "statistics" first gained widespread use to describe the state of society.[12]

During this period, astronomers Marquis de LaPlace and Adolphe Quetelet studied extensively the concept of a statistically derived average value, leading to the concept of the law of errors. The law of errors described departures from the mean values. Neither LaPlace nor Quetelet attempted to describe the distribution of these errors; their efforts were directed at explaining away the errors, usually by placing blame on poor instrumentation.

Actual study of the distribution of the departure of observations from the mean, and its interpretation as genuine variation rather than pure error, was left to mathematician Abraham De Moivre. The development of this particular distribution, which is now known as the normal distribution or bell-shaped curve, marked the beginning of modern-day mathematical statistics and a departure from what is known as classical statistics.

The normal distribution was employed by Carl Friedrich Gauss to fit a large number of astronomical observations to a single value or curve, through developing the methods of least squares. This involved choosing a curve that minimized the sums of the squares of departures of individual measurements from the mean.

The systematic study of departures from the mean was the subject of Francis Galton's work on heredity in the 1860s. His attempt to study whether genius was inherited led him to measure the heights and weights of thousands of fathers and sons. He coined the statistical units of departures of the measurements from their average value a "standard deviation" and demonstrated that the data behaved regularly enough to predict the number of observations within different standard deviations from the mean.[17]

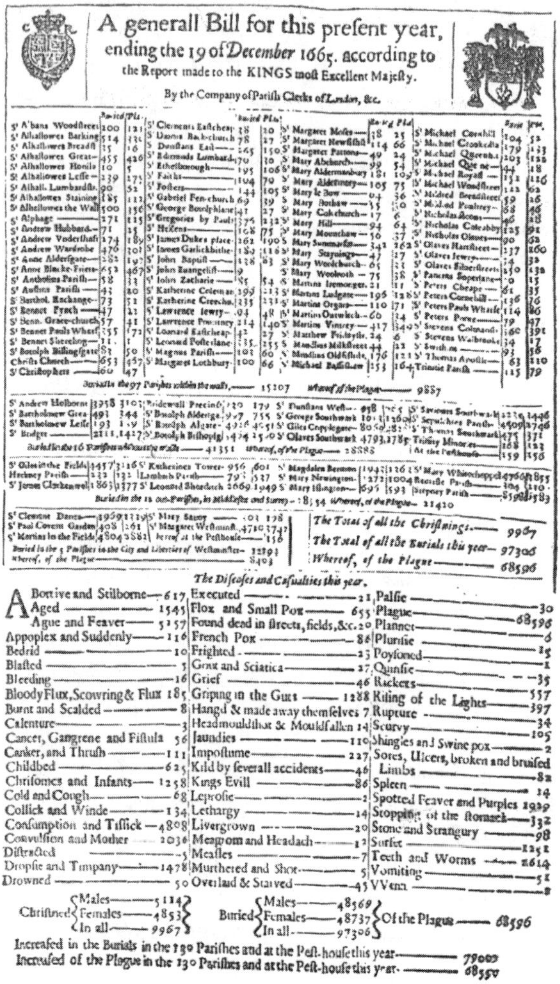

A generall Bill for this present year, ending the 19 of *December* 1665. according to the Report made to the KINGS most Excellent Majesty.

By the Company of Parish Clerks of *London*, &c.

The Total of all the Christnings. — 9967

The Total of all the Burials this year — 97306

Whereof, of the Plague — 68596

The Diseases and Casualties this year.

Abortive and Stilborne	617	Executed	21	Palsie	30
Aged	1545	Flox and Small Pox	655	Plague	68596
Ague and Feaver	5257	Found dead in streets, fields, &c.	20	Plannet	6
Appoplex and Suddenly	116	French Pox	86	Plurisie	15
Bedrid	10	Frighted	23	Poysoned	1
Blasted	5	Gout and Sciatica	27	Quinsie	35
Bleeding	16	Grief	46	Rickets	557
Bloody Flux, Scowring & Flux	185	Griping in the Guts	1288	Rising of the Lights	397
Burnt and Scalded	8	Hang'd & made away themselves	7	Rupture	34
Calenture	3	Headmouldshot & Mouldfallen	14	Scurvy	105
Cancer, Gangrene and Fistula	56	Jaundies	110	Shingles and Swine pox	2
Canker, and Thrush	111	Impostume	227	Sores, Ulcers, broken and bruised Limbs	82
Childbed	625	Kild by severall accidents	46	Spleen	14
Chrisomes and Infants	1258	Kings Evill	86	Spotted Feaver and Purples	1929
Cold and Cough	68	Leprosie	2	Stopping of the stomack	332
Collick and Winde	134	Lethargy	14	Stone and Strangury	98
Consumption and Tissick	4808	Livergrown	20	Surfet	1251
Convulsion and Mother	2036	Meagrom and Headach	12	Teeth and Worms	2614
Distracted	5	Measles	7	Vomiting	51
Dropsie and Timpany	1478	Murthered and Shot	5	Wenn	8
Drowned	50	Overlaid & Starved	45		

Christned: Males — 5114; Females — 4853; In all — 9967

Buried: Males — 48569; Females — 48737; In all — 97306. Of the Plague — 68596

Increased in the Burials in the 130 Parishes and at the Pest-house this year — 79009

Increased of the Plague in the 130 Parishes and at the Pest-house this year — 68590

FIGURE 1.1 John Graunt's "Natural and Political Observations on the Bills of Mortality," 1662.

Galton's study of the heights and weights of parents and children also led to the development of correlation — that is, the mathematical relationship between two variables, such as height and weight. Galton's colleague, Karl Pearson, further explored this concept. Today, we call this the Pearson correlation coefficient, denoted by R^2.

Galton's greatest contribution, however, was his development of regression analysis. His empirical studies on parents and children, as well as his sweet pea breeding experiments, led him to conclude that offspring had the tendency to regress to the mean value of the population at large. Tall parents tended to have slightly shorter children and short parents had slightly taller children; in other words, both sets had a tendency to move, or regress, to the mean value of the population height. Galton called this phenomenon regression to mediocrity (Figure 1.2).[18]

TABLE I.

NUMBER OF ADULT CHILDREN OF VARIOUS STATURES BORN OF 205 MID-PARENTS OF VARIOUS STATURES.
(All Female heights have been multiplied by 1·08).

Heights of the Adult Children.

Heights of the Mid-parents in inches.	Below	62·2	63·2	64·2	65·2	66·2	67·2	68·2	69·2	70·2	71·2	72·2	73·2	Above	Adult Children	Mid-parents	Medians.
Above	..	..	..	..	..	..	..	..	..	..	1	3	..	..	4	5	..
72·5	..	..	..	..	..	..	1	2	1	2	7	2	4	..	19	6	72·2
71·5	..	..	..	..	1	3	4	3	5	10	4	9	2	2	43	11	69·9
70·5	1	..	1	..	1	1	3	12	18	14	7	4	3	3	68	22	69·5
69·5	..	..	1	16	4	17	27	20	33	25	20	11	4	5	183	41	68·9
68·5	1	..	7	11	16	25	31	34	48	21	18	4	3	..	219	49	68·2
67·5	..	3	5	14	15	36	38	28	38	19	11	4	..	..	211	33	67·6
66·5	..	3	3	5	2	17	17	14	13	4	..	..	..	..	78	20	67·2
65·5	1	..	9	5	7	11	11	7	7	5	2	1	..	..	66	12	66·7
64·5	1	1	4	4	1	5	5	..	2	..	..	..	..	..	23	5	65·8
Below	1	..	2	4	1	2	2	1	1	..	..	..	..	..	14	1	..
Totals	5	7	32	59	48	117	138	120	167	99	64	41	17	14	928	205	..
Medians	..	..	66·3	67·8	67·9	67·7	67·9	68·3	68·5	69·0	69·0	70·0	..	..	..	..	..

NOTE.—In calculating the Medians, the entries have been taken as referring to the middle of the squares in which they stand. The reason why the headings run 62·2, 63·2, &c., instead of 62·5, 63·5, &c., is that the observations are unequally distributed between 62 and 63, 63 and 64, &c., there being a strong bias in favour of integral inches. After careful consideration, I concluded that the headings, as adopted, best satisfied the conditions. This inequality was not apparent in the case of the Mid-parents.

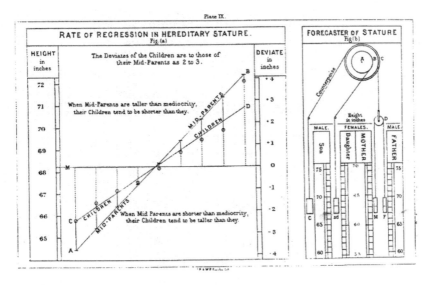

FIGURE 1.2 Francis Galton's regression table, 1885.

Galton's developments were the beginning of today's mathematical statistics. Important developments in statistics in the early 20th century include Ronald Fisher's distinction of small samples vs. the population, the origin of inferential statistics. Fisher's development of the t test (along with William Gosset) and hypothesis testing for small samples led to widespread use of sample statistics. Fisher's other landmark work included developing analysis of variance (ANOVA) methods and Design of Experiments (DOE).

Statistical developments since then have concentrated on nonparametric methods of analysis, which is the study of data that does not follow a normal distribution, exploratory data analysis (EDA), and factor analysis.

Without the concepts of standard deviation and variance, Shewhart would not have been able to develop his ideas on Statistical Process Control (SPC) and identification of random vs. assignable causes. Thus, a mere game of dice 350 years ago leads us to sophisticated analysis of data that is directed toward reduction of variation in our operational processes today.

1.6   QUALITY HALL OF FAME

The preceding section on the history of statistics and the quality movement highlighted several pioneers in laying the foundation for Six Sigma. In tribute to these individuals and others, we recognize them for their significant contributions by induction into the Quality Hall of Fame.

1.6.1   WALTER A. SHEWHART, THE FATHER OF QUALITY (1891–1967)

Walter Shewhart was born in Illinois in 1891 and graduated from the University of Illinois with bachelor's and master's degrees in physics and from the University of California with a doctorate in physics (Figure 1.3).

After graduation, Shewhart joined the Western Electric Company, manufacturers of telephone hardware for Bell Laboratories. The company needed highly reliable transmission wires because the wires were buried underground and were, therefore, expensive to repair. This prompted Bell Labs to invest in research to prevent failures. In a 1924 memo to his superiors, Shewhart proposed the control chart as a method to determine when a process was in a state of statistical control; that is, when the output of the process can be predicted with a high degree of certainty, leading to improved reliability.[19]

Shewhart continued to study the application of statistics to industrial processes and published the landmark "Economic Control of Quality of Manufactured Products" in 1931, which even today is considered a thorough description of the principles of quality control.[20]

Shewhart was considered a gifted mathematician and statistician, as well as a brilliant physicist, studying extensively the methods of measuring the speed of light. His original research and groundbreaking application of statistics to industry, as well as his mentorship of subsequent quality giants, earns him a special place in the Quality Hall of Fame.

1.6.2   W. EDWARDS DEMING (1900–1993)

W. Edwards Deming has probably had the greatest role in shaping the quality profession.[21] Building on the groundbreaking methods of his mentor, Walter Shewhart, Deming's principles of profound knowledge, common and assignable variation and process management, and his Plan–Do–Check–Act (PDCA) continuous improvement approach permeate the majority of quality improvement practices to this day.

FIGURE 1.3   Walter A. Shewhart.

FIGURE 1.4   W. Edwards Deming.

Dr. Deming was born in 1900 and graduated from Yale University with a doctorate in mathematical physics (Figure 1.4). While working summers at the Western Electric Hawthorne Lighting Works, he met Walter Shewhart and quickly gained an appreciation for Shewhart's quality control techniques. In 1938, Deming, while working for the U.S. Department of Agriculture (USDA), invited Shewhart to lecture at the USDA graduate school. From these lectures, Deming published "Statistical Methods from the Viewpoint of Quality Control."[22]

During World War II, Deming taught defense contractors at Stanford University SPC methods for use in manufacturing war materials. By 1945, more than 30,000 engineers and technicians had been given basic SPC training.[23] Although the war effort took advantage of Shewhart and Deming's techniques, U.S. business managers made no attempt to understand and apply the techniques to their own manufacturing.

After WWII, Deming traveled to Japan to collect data for the U.S. Department of War. During these visits he became close with Japanese statisticians from the Japanese Union of Scientists and Engineers (JUSE). They were interested in Deming's SPC methods to help with the reconstruction of war-ravaged Japan.

In June 1950, at the invitation of the head of the JUSE, Dr. Deming taught statistical methods to a group of 230 students in Tokyo. After those classes, he realized that only the top management in a company could really make a difference in the quality of products. He persuaded Dr. Ishiro Ishikawa, the head of Kei-dan-ran, an association of Japanese executives, to meet with him. After discussions with Ishikawa, he presented his ideas to the association members, who began applying the techniques with notable success.[24]

Working together, Dr. Deming and the Japanese industrialists began to understand the need to continually redesign and refine their processes and products. This methodology became known as the Deming cycle, what we know today as the PDCA continuous improvement cycle.

Six Sigma is based largely on the work of Shewhart and Deming. The focus on statistical measures to reduce variation is at the core of both their methods and Six

Sigma. Although it may seem obvious today that these are the best methods to improve process and product quality, these two individuals developed and pioneered these concepts, as well as the tools and techniques necessary to apply them in real life. Thus, W. Edwards Deming joins Walter A. Shewhart in earning a special place in the Quality Hall of Fame.

1.6.3 JOSEPH M. JURAN (1904–)

Like his colleague, Dr. Deming, Joseph Moses Juran worked at the Western Electric Hawthorne Lightning Works in the 1920s, right out of college. Juran graduated from the University of Minnesota with an electrical engineering degree in 1924 and from Loyola in 1935 with a law degree. At the Hawthorne plant, Juran met Walter Shewhart, who was consulting there at the time. Juran was one of two engineers chosen to work with Shewhart in applying statistical methods to business operations.[25]

Similar to Deming, Juran was well known in Japan for his contributions to Total Quality Control after WWII. One of Juran's most significant contributions to the quality field is the widely used Pareto principle. In 1937, he took the idea of Vilfredo Pareto, an Italian economist in the late 1800s, that 20% of the people hold 80% of the wealth and applied it to the study of business operations data. Juran called it separating the vital few from the trivial many and applied the concept to defect analysis. He noted that efforts should be concentrated on the vital few and called it the Pareto principle.[25]

Juran is also known for the Juran Quality Trilogy process, which he believes is necessary to "break through" to new levels of quality. The Juran Quality Trilogy is:

1. Quality planning: The process for developing quality goals and a plan to meet them. It involves understanding customer requirements and creating the process for meeting the requirements (targets).
2. Quality control: Ongoing measurement of process quality through measuring current performance and the gap between performance and targets. Action steps are taken to close the gaps and keep the process in statistical control.
3. Quality improvement: Identifying improvement projects based on quality control measurements, organizing improvement teams, diagnosing and proposing solutions to the problems, proving their effectiveness, and implementing them, with quality control measures in place for the new process.

Juran uses a project orientation approach to improvement. His philosophy is that improvements are accomplished step-by-step, project-by-project. He is also known for his humanistic perspective to quality improvement, a trait shared by Deming. He advocated the use of quality circles as improvement teams, an innovative concept at the time, because the workers rather than their managers were involved in improving the quality of their process. Today, the practice is widespread, although the term quality circle is rarely used.[25]

Juran was the first quality expert to recognize money as the language of management and compared quality planning to budgeting, quality control to cost control, and quality improvement to cost reduction. A significant concept in Six Sigma, the cost of poor quality (COPQ), originated with Juran's analogies of the cost of quality with those of financial planning.

For his many contributions to the quality management field, Joseph M. Juran joins his colleagues, Walter Shewhart and Dr. Deming, in the Six Sigma Quality Hall of Fame.

1.6.4 PHILIP R. CROSBY[26] (1926–2001)

Philip R. Crosby gained national recognition in 1979 with his bestselling book *Quality Is Free*. Crosby's early experience in the quality field was at International Telegraph and Telephone (ITT), where he was corporate quality manager for 14 years.

Crosby was born in West Virginia and served in both WWII and the Korean War, graduating from Western Reserve University between his military stints. When he joined defense contractor Crosley in 1952 his first assignment was quality manager for the first Pershing missile program. He also worked for Martin-Marietta for 8 years before moving to ITT in 1965.

While at ITT, Crosby published *Quality Is Free*, in which he stressed that traditional quality control methods encouraged failures because there is some acceptable level of deviation from requirements. He estimated this costs a typical service company 35% of operating expenses. Crosby's message was that only through prevention could a company achieve high quality at a low cost. He believed that achieving high quality resulted from a management-led cultural transformation that emphasized prevention rather than a statistics-based quality control program for detecting defects.[27]

Crosby is best known for his Zero Defects and Do It Right The First Time concepts. Similar to Six Sigma, his Zero Defects concept is not always meant to be taken literally, but rather used as an approach toward quality improvement.

Similar to Deming, he believed that poor quality is not the fault of production line workers, but because management does not set the proper tone for preventing quality problems. He advocated a top-down approach to quality, spread throughout the company by a core team of quality experts to result in a cultural transformation. However, he believed that responsibility for quality improvement is entirely owned by senior management. Crosby was also one of the first quality experts to stress that quality improvement is a continuous process rather than a program.

The success of Crosby's book led him to leave ITT in 1979 and form Philip Crosby Associates, where he developed a Quality College to provide training in his concepts. Crosby developed 14 steps toward quality improvement, designed to provide the structure and the process for achieving zero defects. His Four Absolutes of Quality Management, as described below, are required for the cultural transformation necessary for high quality:[28]

1. Quality is defined as conformance to requirements rather than goodness of fit.
2. Prevention, rather than inspection for defects, is the only way to achieve quality.

3. Zero defects, rather than "that's close enough," is the appropriate performance standard.
4. The true measure of quality is the price of nonconformance (similar to Six Sigma's COPQ), rather than a percentage quality level measure.

Crosby's innovative contributions in preventative quality concepts and top-down management support, concepts that we often take for granted today, have earned him a place in the Quality Hall of Fame.

1.6.5 SHIEGO SHINGO (1909–1990)

Shiego Shingo is not as well known as some of the other quality experts; however, his contributions to quality improvement, and more notably Lean Production, are considerable. Shingo's most lasting contributions to the quality field are the concepts of mistake proofing for zero defects and Single-Minute Exchange of Dies (SMED), which drastically reduces manufacturing cycle time and allows a JIT manufacturing system.

Born in Saga City, Japan, Shiego graduated in mechanical engineering from Yamanashi Technical College in 1930. After college, he worked for the Taipei Railway Factory in Taiwan and subsequently became a management consultant for the Japan Management Association. During this time, he applied SPC techniques in a wide variety of applications and companies, including Toyota Motor Company and Mitsubishi Heavy Industries.[29]

In 1959, Shiego founded his own consulting firm, where he trained more than 7,000 people in industrial engineering and quality improvement techniques. It was during this period that he, along with Taiichi Ohno, developed the concepts of mistake proofing and SMED.

Mistake proofing, also called Poka-Yoke (pronounced POH-kuh YOU-kay), parallels Crosby's concept of Zero Defects. Like Crosby, Shingo came to realize that traditional quality control techniques would not in themselves produce high quality. He believed high quality is achieved only through stopping the defect at the source; this concept was translated operationally by having production workers detect the errors at their source and stop the production process until the cause of the error was understood and corrected.

Another of Shingo's important contributions is the SMED system. This system involves reducing the time it takes to changeover from producing different products on the same manufacturing line. It is accomplished through using common set-up elements when possible and simplifying the exchange of dies when necessary. This results in significantly reduced manufacturing cycle time and allows small-batch production using a JIT system.

Shingo's techniques are the basis of his Zero Quality Control principles, which revolutionized the manufacture of goods through reducing cycle times and improving quality at significantly lower costs. His success is evidenced by the annual Shingo Prize, which recognizes companies for achieving significant improvements in quality and productivity. For his lasting contributions to quality improvement and Lean Manufacturing, Shingo has earned a place in the Quality Hall of Fame.

1.6.6 TAIICHI OHNO (1912–1990)

Taiichi Ohno is best known for his development of the Toyota Production System (TPS), which is the basis for Lean Manufacturing. Borrowing concepts of continuous-flow manufacturing from Henry Ford's mass production system, TPS incorporates the concepts of elimination of all wastes, preventive quality, JIT inventory, and small-lot production.

Ohno joined the Toyota Motor Company after WWII as a production engineer. On a trip to Detroit after the war, Ohno noted the number of specialists involved in making a car was almost equal to the number of production workers that added value to the car. He also noted other wasteful practices such wasted effort, material, and time.[30]

Back in Japan, Ohno began experimenting with producing cars through increased reliance on production workers and less on specialists. Because the demand for vehicles was low, Ohno needed to find a way to mix production among several different auto lines in an efficient way. He worked with Shiego Shingo to develop the SMED system, which allowed quick turnover of production lines and small-lot production. From this evolved the idea of JIT supply practices. He also worked with Shingo to implement mistake proofing and stopping the production line at the source when a defect was detected.[30]

When Ohno again traveled to Detroit in 1956, he visited a supermarket, where he noted how shoppers picked from an array of goods, which were then replenished by the stock clerks. Ohno applied this idea to the factory, where each production line served as a supermarket for the succeeding line — items were selected and replaced as needed from one line to the next. Ohno called this the pull system of production, which is a principal component of Lean Manufacturing.

In 1947, Toyota produced just 100,000 vehicles; in 2000, it produced 5.8 million, making it the third largest producers of vehicles in the world. Toyota's automobiles consistently rank at or near the top in customer satisfaction ratings and the company is well known for its innovative technology involving car safety and emissions control systems. It is doubtful that this would be the case without Ohno's significant contributions. From TPS, Lean Manufacturing evolved and has been successfully implemented in thousands of companies worldwide, resulting in economic gains and a higher standard of living for many people. For these reasons, Ohno is a well-deserved candidate for the Quality Hall of Fame.

REFERENCES

1. "Making Customers Feel Six Sigma Quality," http://www.ge.com/sixsigma/making-customers.html, 2001.
2. McGuire, Miles, Ed., Cowboy quality, *Quality Progress*, October 1999, p. 28.
3. McGuire, Miles, Ed., Cowboy quality, *Quality Progress*, October 1999, p. 30.
4. McGuire, Miles, Ed., Cowboy quality, *Quality Progress*, October 1999, p. 31.
5. Eckes, G., *Making Six Sigma Last: Managing the Balance between Cultural and Technical Change*, John Wiley & Sons, New York, 2001, p. 207.

6. Lazzaro, Victor, Ed., *Systems and Procedures: A Handbook for Business and Industry*, Prentice-Hall, Englewood Cliffs, NJ, 1959, p. 128.
7. Lazzaro, Victor, Ed., *Systems and Procedures: A Handbook for Business and Industry*, Prentice-Hall, Englewood Cliffs, NJ, 1959, p. 157.
8. "Transferring Lean Manufacturing to Small Manufacturers: The Role of NIST-MEP," http://www.sbaer.uca.edu/Docs/proceedingsIII.htm, 2001.
9. "Creating a Future State," http://www.Lean.org/Lean/Community/Resources/thinkers_start.cfm, 2001.
10. Womack, James P., Jones, Daniel T., and Roos, Daniel, *The Machine that Changed the World: The Story of Lean Production*, HarperPerennial, New York, 1990, p. 86.
11. http://www.nist.gov/public_affairs/factsheet/baldfaqs.htm.
12. Porter, T., *The Rise of Statistical Thinking 1820–1900*, 1st ed., Princeton University Press, Princeton, NJ, 1986.
13. For example, the chances of winning a lottery in which six numbers are randomly chosen are 1 in 13,983,816.
14. http://www.anselm.edu/homepage/jpitoch/biostatistime.html, 2001.
15. http://www.ac.wwu.edu/stephan/Graunt/bills.html, 2001.
16. http://www.mrs.umm.edu/sungurea/introstat/history/w98/galton.html.
17. Zikmund, William G., *Business Research Methods*, 4th ed., Dryden Press, New York, 1994, p. 557.
18. http://asq.org/join/about/history/shewhart.html, 2001.
19. Wheeler, Donald J., and Chambers, David S., *Understanding Statistical Process Control,* 2nd ed., SPC Press, Knoxville, TN, 1992, p. 6.
20. I attended Dr. Deming's 4-day seminar in 1992, the year before his death. He was in a wheelchair and spoke very softly. The effort involved in lecturing at this point in his life was considerable; however, there was a spark in his eye that belied the effort. His apparent love for the subject was contagious among the audience, resulting in one of the more memorable learning experiences of my career.
21. Wheeler, Donald J., and Chambers, David S., *Understanding Statistical Process Control,* 2nd ed., SPC Press, Knoxville, TN, 1992, p. 8.
22. From this group, the American Society of Quality Control, the forerunner of the American Society for Quality (ASQ), was founded in 1946.
23. http://www.deming.eng.clemson.edu/pub/psci/files/3expert.txt.
24. http://www.stanford.edu/class/msande269/six_scholars_comparisons.html.
25. I met Crosby at a Quality and Productivity Research conference in 1993, where he was a keynote speaker. I had the opportunity to discuss with him the TQM approach, which I was involved in implementing at American Express at the time, and how his Zero Defects concepts compared to TQM. Mr. Crosby was a genuinely nice, enthusiastic, and down-to-earth person, who was vocal in his beliefs of preventive quality concepts. Philip Crosby died while this book was in development, and with his passing we have lost one of the true quality giants of the 20th century.
26. Johnson, Kristen, Ed., Philip B. Crosby's mark on quality, *Quality Progress*, October 2001, p. 25.
27. Crosby, Philip B., *Quality Is Free: The Art of Making Quality Certain*, McGraw-Hill, New York, 1979.
28. Shingo, Shiego, *A Study of the Toyota Production System from an Industrial Engineering Viewpoint*, Productivity Press, Portland, OR, 1989.
29. Ohno, Taiichi, *Toyota Production System — Beyond Large-Scale Production*, Productivity Press, Portland, OR, 1988.

2 Organizational Success Factors

2.1 LEADERSHIP MATTERS

Strong commitment, leadership, and strategic involvement from Six Sigma's early adopters have proven to be key factors of its success. Bob Galvin, chairman of Motorola, committed to the program in 1987; Larry Bossidy of Allied Signal adopted it in 1994; and Jack Welch of GE followed suit in 1995.

Galvin's commitment to the Six Sigma approach was so strong that he delivered the keynote speech at the Juran Institute in 1992 addressing the "quality" issue from the perspective of a CEO.[1] By doing so, Galvin clearly and publicly demonstrated his commitment to the Six Sigma methodology and endorsed it as a strategic initiative. Although it took 10 years for Six Sigma to be fully integrated into Motorola's culture, Motorola clearly reaped the positive business results from the process from the beginning.[2] Quotes from the annual report include a "five fold growth in sales with nearly 20% annual increase in profits," "cumulative savings based on Six Sigma efforts at $14 billion," and "compounded annual stock price gain of 21%."

Such results would have been unlikely from a short-term commitment to such a substantial change initiative. As the pioneer in implementing Six Sigma, Motorola encountered several hurdles and learned key lessons along the way. Much of the knowledge gained has been incorporated into subsequent implementation models used by Allied Signal, GE, and others. By reaping the benefits of Motorola's hindsight and following the best practices from the experiences of others, an organization can significantly reduce the length of implementation. Many companies begin experiencing results from their Six Sigma efforts within 6 to 12 months from initiation of their first projects. However, it usually takes 3 to 5 years before the Six Sigma customer-centric and continuous improvement processes are an integral part of the organization.

Larry Bossidy, who took over as CEO of Allied Signal in 1991, experienced tremendous success with Six Sigma. Both at Allied Signal and, starting in 1999, at Honeywell, Bossidy succeeded in developing a culture that embraces Six Sigma. A testament to this is that his leaders have come to view Six Sigma as "more than just numbers — it's a statement of our determination to pursue a standard of excellence using every tool at our disposal and never hesitating to reinvent the way we do things."[3] Six Sigma continues to be a tour de force within Honeywell. The initiative is currently called "Six Sigma Plus" and is defined as "one of the most potent strategies ever developed to accelerate improvements in processes, products, and services, and to radically reduce manufacturing and/or administrative costs and improve quality."[4] Furthermore, Six Sigma Plus is seen as the "principal engine for driving growth and productivity across all businesses."[4]

Jack Welch was initially reluctant to embrace Six Sigma. But Bossidy, his close friend and former colleague from GE, persuaded Welch that it was an initiative well worth the investment. In 1995, Welch committed to the idea of launching GE's Six Sigma effort based on his belief that a strategic focus on quality improvement could provide significant business results. In Welch's estimation, the cost to GE of *not* implementing Six Sigma would be the loss of realizing annual savings of approximately 10 to 15% of its revenue ($8 to 10 billion). GE expected to gain this back and more within 5 to 7 years. In January 1996, Welch launched the Six Sigma effort at a GE executive meeting; he told his managers he expected that it would transform GE in a way that appealed to both the shareholders and the staff. Welch proclaimed that GE's commitment to Six Sigma would be "the biggest opportunity for growth, increased profitability, and individual employee satisfaction in the history of the company."[5] Next, Welch appointed one of his senior staff members to be the Six Sigma Champion and to have oversight responsibility for the initiative.

Although Welch's tremendous leadership abilities made it look easy to successfully launch the change effort, most business managers know the challenges in implementing a new initiative such as Six Sigma. The key lesson learned from Galvin, Welch, and Bossidy is that a CEO's commitment to Six Sigma's quality process is imperative for gaining and sustaining momentum, and ultimately achieving success. This commitment often stems from the business need to survive among global competition, technology advancements, and demanding customers; and like GE, companies must ask, What is the price of not implementing?

2.2 SIX SIGMA AS A STRATEGIC INITIATIVE

The challenge for any business leader committed to implementing change is to first build a business case for its implementation. Because Six Sigma requires a long-term mentality, it needs to be positioned first as a strategic initiative and then be linked to operational goals. Its implementation needs to be tied to corporate goals and organizational leaders must be able to clearly link Six Sigma to the needs of the company, such as increased profits through lower costs and higher customer loyalty.

One of the primary reasons for leadership commitment to Six Sigma is that most projects span across several departmental, functional, and vertical lines within the organization's hierarchy. Companies whose leaders place strategic importance on Six Sigma find that managers responsible for functional or divisional units demonstrate support for projects by removing organizational barriers and making significant changes to facilitate improvements. Any change initiative involves resistance by those with a stake in the existing environment, which is most likely the individuals who have worked hard and achieved success as managers. If Six Sigma projects are linked to a strategic imperative, the business managers may more readily resolve cross-functional conflicts, understand the need for process improvements, and be more willing to adapt to necessary changes. The business rationale behind Motorola, Allied Signal, and GE's decision was the imperative to improve product quality — or else. Clearly, they succeeded. Below are excerpts from the annual reports of corporations that have positioned Six Sigma as a strategic competitive weapon:

> Ford's 2000 annual report: "In the past year we launched Consumer Driven 6-Sigma, a scientific, data-driven process to uncover the root cause of customer concerns and drive out defects ... saving the company $52 million."[6]
>
> Honeywell's 1999 annual report "Delighting customers and accelerating growth completes the picture. When we are more efficient and improve workflow throughout every function in the company, we provide tremendous added value to our customers."[7]
>
> Johnson Controls' 2000 annual report: "Six Sigma is transforming our quality and cost improvement effort from one of evolutionary progress to revolutionary results. Applying Six Sigma across the organization will help [us reach] stronger competitive positions, increased market shares and greater shareholder value."[8]
>
> Dupont's 2000 annual report: "Our three strategic pathways of Integrated Science, Knowledge Intensity and Six Sigma Productivity are propelling us toward our goal of sustainable growth."[9]

These examples illustrate the strategic relevance organizational leaders place on Six Sigma and the integral part it plays in the company's business strategy and operations. These leaders have established a clear direction, provided rationale for the direction, and use this strategic focus to influence others to support and act in congruence with the organization's goals.

2.3   INTERNAL COMMUNICATION STRATEGY AND TACTICS

An organization's CEO must be philosophically committed to the Six Sigma initiative in the sense of believing that continuous process improvement is a fundamental way of life in modern businesses. Such philosophical commitment is necessary to sustain the organizational commitment to Six Sigma and provides the foundation for an effective communication strategy. Research has consistently shown that top management's attitude influences the behavior of other managers in an organization.[10] The leader who chooses to champion Six Sigma by consistently demonstrating enthusiastic support, staying involved in project reviews, publicizing the project teams' results, and rewarding participants involved in the implementation has a much greater probability of a successful implementation. Such leaders realize that effective communication is key for building and sustaining organizational commitment to Six Sigma.

An effective communication strategy for implementing Six Sigma includes two components — the message itself and the way the message is delivered:

1. First, the company must educate employees on the tactical issues relative to implementing Six Sigma by explaining the who, what, why, where, when, and how's. The goal of this phase is to ensure that employees have a background understanding of the initiative as well as some of the specific facts and goals related to its implementation.

2. The second component focuses on *how* the company effectively communicates the message to all employees in a way that increases their understanding and commitment to the Six Sigma implementation. This is accomplished using both formal and informal communication plans. The formal plan is used to launch the Six Sigma effort, formalize it as a strategic initiative, and keep employees informed of its progress through executive briefings and other announcements. The informal communication plan is to ensure regular communication about Six Sigma and its impact on the organization using various channels, including newsletters, bulletins (e-mail updates on project status and results), and regularly held update meetings with managers and staff.

The tactical component of the formal communication plan is designed to ensure that all employees understand the Six Sigma vision and objectives strategy, addressing the following questions:

- What is the Six Sigma methodology?
- Why is this initiative important to the organization now?
- What are the goals of implementing Six Sigma within the organization?
- Who will be affected by this initiative?
- Who will be involved in implementing the initiative?
- Where in the organization will the Six Sigma efforts be focused?
- When will the initiative begin?
- How long is it expected to last?
- How will employees learn more about how Six Sigma impacts their job?
- How and when will employees become trained to implement Six Sigma projects?

This list is not intended to be all inclusive. An organization may elect to include additional questions or address other relevant issues as they plan to launch Six Sigma.

2.4 FORMALLY LAUNCHING SIX SIGMA

After the tactical message is developed and reviewed with the executive team, the next phase is to create a communication plan for the Six Sigma launch. This phase can take several different approaches; however, it is strongly recommended that the CEO deliver the message and that he or she do so with fervor and enthusiasm. If the CEO is reluctant to do this in the initial launch, there is reason to be concerned for the entire effort.

Some CEOs have chosen to introduce Six Sigma at formal management meetings or at the company's annual employee meeting where the strategic plan is discussed. The primary goal of the introduction is to inform leaders and staff members of the CEO's commitment to Six Sigma and to create a sense of urgency around it. The reason for creating urgency is to quickly elevate Six Sigma's importance to all employees with the message that the initiative is required for survival in the competitive marketplace. Without this sense of urgency backed by quick action, employees may soon disregard its importance.

During the formal launch, the CEO needs to communicate a concise vision of how Six Sigma is expected to benefit the employees, the organization, and its shareholders. A few of these benefits are listed as follows:

- To the employees: career growth, job satisfaction, and an opportunity to contribute to the organization in new ways
- To the organization: improve processes, cross-functional problem solving, and increase customer satisfaction
- To the shareholders: economic performance through organizational improvements

In addition to formal communications, the company needs to use various informal methods for ongoing communication. Examples include newsletters, news flashes, videos, intranet sites, and even CEO spotlight mini presentations. The CEO's continued involvement in the communication effort through words and visible action demonstrates the seriousness and strategic importance driving the initiative. If the CEO has a limited role in the communication process, then expect limited action toward the Six Sigma effort. On the other hand, if the CEO makes Six Sigma a priority, employees concerned about their job security would be wise to do the same. Jack Welch made it clear that anyone who wanted to be promoted at GE had to complete Six Sigma Black or Green Belt training.[11] Other CEOs may not be as direct, yet will have similar expectations for anyone interested in assuming additional responsibility within the company.

Company managers, in particular, middle mangers, are the backbone of the informal communication efforts; discussions involving Six Sigma should be viewed as an extension of managerial processes, not just an event. To the extent that leaders and managers continuously communicate with their colleagues and staff members about Six Sigma, and link the business strategy with daily operations, the effort will begin to take hold as an element of the organization's regular business practices and culture more rapidly and effectively.

Managers and leaders are encouraged to discuss the realties of implementing Six Sigma with their staff and encourage face-to-face and two-way communication about the reasons for change and the impact it will have on others — the positive and the challenging. They must also display a willingness to address challenging questions that arise from inquisitive colleagues, subordinates, and internal skeptics. Such openness helps to uncover resistance to change, which is the first step in overcoming it.

2.5 ORGANIZATIONAL STRUCTURE

Migrating to a culture that supports Six Sigma requires structural and transformational change. The foundation for the change is established through an organization's identification of the Six Sigma business strategy. The initial changes include forming a small group to lead the change effort. This is followed by training to teach employees new skills for implementing Six Sigma.

2.5.1 LEADING THE EFFORT

When launching Six Sigma, the CEO should appoint a small group of senior executives, representing various functions and business units, to provide leadership and oversight to the Six Sigma implementation. This group acts as an extension of the CEO's leadership, removing operational and organizational barriers to success while providing leadership to the Six Sigma teams. In smaller organizations a single executive may provide this role. The name given to this senior executive group responsible for leading the Six Sigma effort varies from company to company; some call it a guiding coalition, others a steering committee or Six Sigma advisory board. Regardless of the name, the function of this group is to make decisions regarding allocation of resources, timing, scope, and method of implementation. Other roles and responsibilities of this group include:

- Building commitment to the Six Sigma implementation among all employees, including the business unit managers responsible for Six Sigma projects in their areas.
- Assisting managers in implementing recommended changes from the Six Sigma teams and fielding any resistance to change as a result of projects.
- Working with the Six Sigma teams to establish processes for planning and executing Six Sigma projects by setting internal standards for project implementation.
- Providing direction by formalizing and conducting management communication meetings regarding Six Sigma implementation and progress.

Finally, they need to be the consummate champions of Six Sigma in their communications and contact with fellow colleagues and staff.

2.5.2 SELECTION CRITERIA FOR SIX SIGMA ROLES

The selection criteria for key roles need to be part of the early implementation planning efforts. Developing clear criteria for Master Black Belts, Black Belts, and Green Belts is critical. Organizations should use the same level of scrutiny in selecting individuals for Six Sigma roles as they do in hiring new employees. Several criteria apply to each of the Six Sigma belt levels. The higher the belt level, the greater degree of ability and experience required. The recommended selection criteria, in order of priority, are listed below:

1. *Intrinsic enthusiasm.* Individuals with natural motivation will be able to serve as informal role models and cheerleaders for the Six Sigma initiative.

This is particularly important during the early stages, when there is more skepticism toward the effort, and is also necessary during the middle stages of the implementation, when enthusiasm and continued success are necessary to keep people motivated and committed.

2. *Organizational credibility.* One must be respected by management within the organization and demonstrate professionalism and integrity.

3. *Technical competence.* Superior analytical skills are necessary for the role of Black Belt and to a lesser degree for Green Belts. The role of Master Black Belt requires many years of experience in applying statistical concepts to business operations.

4. *Process orientation.* Individuals must have an understanding of how a set of tasks comprise a process and that effective processes contribute to work efficiency. These individuals must also be able to establish, maintain, and improve internal processes.

5. *Goal orientation.* One must have a demonstrated track record of setting and achieving organizational goals.

6. *Effective communication skills.* The employee must be able to write clearly and concisely, conduct management presentations, and facilitate group meetings, as well as work effectively with colleagues at all organizational levels.

7. *Natural or emerging leadership skills.* The individual must have demonstrated an ability to gain the respect of others and to motivate others to take action.

8. *Project management and organizational skills.* One must demonstrate an ability to plan, execute, and manage projects utilizing project management practices.

2.6 SIX SIGMA TRAINING PLAN

Ongoing training and development are integral parts of a Six Sigma implementation. Because employees are expected to learn and demonstrate new skills, some level of training for all levels of the organization is recommended. The first set of classes are geared toward the leadership group, with the objective not only of clarifying what Six Sigma is, but more importantly understanding the roles and responsibilities the leadership group is expected to demonstrate. Next comes Six Sigma technical training for the key Six Sigma project roles. The training is targeted to each Six Sigma role and level of responsibility.

2.6.1 MASTER BLACK BELT TRAINING

The overwhelming majority of candidates for Master Black Belt have previously completed the Black Belt training and have many years of experience in applying statistical tools to improve quality and the productivity of an organization. Demonstrated leadership and mentoring skills are also a prerequisite of this role. A Master Black Belt is certified to teach Champions, Black Belts, and Green Belts on Six Sigma methodology, tools, and applications. They coach Black Belts on their respective Six Sigma project implementations and develop and oversee standards to ensure that those implementations are consistent among the various project teams. They also communicate, reinforce, and motivate others to embrace the Six Sigma vision, facilitate in removing organizational stumbling blocks, and participate in selecting Six Sigma projects.

2.6.2 BLACK BELT TRAINING

This training prepares a managerial or mid-level professional to lead Six Sigma projects. This 4-week training is usually delivered over a period of 3 to 4 months and uses multiple experiential, hands-on, and interactive teaching methodologies. In-depth coverage of how to utilize the DMAIC methodology, which is the foundation for the Six Sigma process, is provided. Six Sigma statistical tools and techniques, along with problem-solving approaches are included within this curriculum. Typically, the participant brings to the training a company-approved project that will be the focus of applying the knowledge, processes, tools, and techniques learned throughout the Black Belt training.

The Black Belts lead Six Sigma projects using the tools and techniques learned in training. Generally, Black Belts are 100% dedicated to working on the Six Sigma projects because they may have anywhere from three to five Green Belts assigned to them. The Black Belts provide expertise, direction, and ongoing coaching to the Green Belts and their respective Six Sigma teams. Some companies use the model of the Master Black Belt advising the Black Belt, who, in turn, supports the Green Belts in conducting projects. Other companies do not use Green Belts, using only Black Belts to lead the Six Sigma projects. Black Belts also communicate and support the Six Sigma vision and are typically the designated project managers for the Six Sigma projects. They may also prepare and deliver presentations to management on project status and results.

2.6.3 GREEN BELT TRAINING

This training typically is an 8- to 10-day course delivered in two sessions over the course of a month. The training focuses on learning a combination of the DMAIC approach and other less technical statistical and problem-solving tools included in the Black Belt training. Green Belts also receive training on effectively managing the project and the overall process.

Green Belts act as project leaders under the supervision of a Black Belt. Generally, Green Belts dedicate anywhere from 10 to 50% of their time to Six Sigma projects while continuing with their regular job responsibilities. Green Belts are more directly involved with the internal process owners and in collecting and analyzing the project data. Green Belts may also participate in presentations to management on project status and results.

2.6.4 TEAM TRAINING

It is important to acknowledge the other members of the Six Sigma projects who do not attend the specialized Black Belt or Green Belt training. These individuals, ranging from process owners to subject matter experts, provide valuable insight into specific job functions or work processes that are the focus of a Six Sigma project. In selecting the team, it is important to include employees whose collective wisdom represents the critical aspects of a process, including, but not necessarily limited to, the following:

- Understand the customer requirements
- Understand the formal and informal aspects of the process
- Represent different perspectives of customer needs based on their level in the management hierarchy
- Have access to critical data that will be important for collection and analysis

Specialized training, including both technical and team skills training, should be provided to team members to prepare them to work effectively on their Six Sigma project teams. Types of skills training that contribute to effective teamwork and problem solving include:

1. *Team effectiveness.* Learn the characteristics of an effective team, understand the stages of team development, understand the roles and responsibilities of team members, and identify action plans that team members can implement to improve their level of contribution to the team.
2. *Problem solving and decision making.* Distinguish a problem's symptom from its root causes, understand how to diagnose the root cause of a problem, avoid quick conclusions, and learn brainstorming and problem-solving techniques.
3. *Effective communication.* Understand different styles of communication, identify one's own personal communication preference, learn actions that trigger breakdowns in communication and how to prevent them, learn team facilitation skills, and identify action plans for improving one's own communication effectiveness.

In addition to the above soft-skills training, team members need technical training that provides an overview of Six Sigma and the basic DMAIC methodology. Training team members also lays the foundation for eventual integration of the Six Sigma approach into all of the organization's key processes.

2.6.5 CHAMPION TRAINING

This training is geared for middle managers to the top leaders of an organization. At a high level, the training accomplishes three main objectives: (1) it provides an overview of Six Sigma, its core philosophy, tools and techniques, success stories, and lessons learned; (2) it reinforces Six Sigma as a top-down business strategy that requires strong leadership support and provides tangible examples of how such leadership can be demonstrated; and (3) it clarifies the roles and responsibilities of Champions to support the Six Sigma process.

The Champion training uses an interactive process to determine the readiness of the organization to implement Six Sigma from both a business and cultural perspective. On the business side, the participants determine the extent to which the organization faces repeated issues with respect to quality, work processes, and meeting customers' needs. Then a quick analysis of the costs of these problems is calculated and put in the perspective of needing to implement Six Sigma. From a cultural perspective, the participants evaluate the aspects of their current organizational culture that may hinder a successful Six Sigma implementation and take action to remove these barriers.

The primary roles and responsibilities of a Champion are to reinforce the CEO's vision of Six Sigma, empower others in support of the Six Sigma objectives, and take ownership of projects they champion, including changing systems or processes that are barriers to successful completion of projects. Champions are involved in establishing criteria for Six Sigma project selection and for approving projects at their completion. Participants leave the Champion training with a strong understanding of not only what Six Sigma is and how it can improve the organization, but also what they as leaders can do to accelerate acceptance and increase the probability of success.

2.7 PROJECT SELECTION

Six Sigma success is not an accident. Implementing Six Sigma in any organization requires significant commitment and effort from the organization's leaders, managers, and other staff; however, its overall success depends greatly on the cumulative successes of the individual projects. For this reason an organization must use a disciplined approach to identify and select the Six Sigma projects that will make a difference. The selection process includes the following activities:

1. Identify who needs to be involved in the selection of Six Sigma projects.
2. Link the projects to the organization's strategic issues.
3. Develop specific selection criteria against which all potential projects will be assessed.
4. Determine a selection decision process.

2.7.1 IDENTIFY DECISION MAKERS

Who needs to be part of the project selection process? The key issue here is ensuring executive-level support for the Six Sigma projects selected. It follows, then, that executives are part of the discussions in identifying potential projects. Decision makers should include the Six Sigma advisory board members, as well as other key players, such as Master Black Belts and Black Belts. Their role is to help the executive team define the scope of the Six Sigma projects and to offer their experience in determining the feasibility and manageability of the projects under consideration.

2.7.2 Link Projects to Strategic Issues

All Six Sigma projects must have strategic relevance and be linked to core business issues. This starts with executives identifying the top business issues the organization is facing relative to competitive pressures discussed in Section 2.1. Chances are the current business issues relate to either the execution of the business strategy or core internal processes that are ineffectively executed and have a negative impact on the organization and its customers.

2.7.3 Develop Selection Criteria

After determining that a project has strategic relevance, the next step is to develop additional criteria by which the project will be judged and ultimately accepted as a Six Sigma project. Below are several suggested criteria that have proven successful. There may be additional criteria specific to a company's requirements:

- Manageable scope — Six Sigma projects must have clearly defined boundaries and deliverables. They should be specific enough in scope so that a Six Sigma team can complete the project in the allotted time, which is usually 4 to 6 months.
- Ongoing process issue — Projects should address an ongoing process issue through which value is gained if resolved; avoid selecting projects related to problematic recent incidents that would not qualify as being a process.
- Measurable outcomes — Projects under consideration should have preliminary measurable outcomes associated with them. As for identifying potential measurable outcomes, consider the following examples: potential to reduce operational costs, increase product or service quality, secure new customers, increase percentage of customers retained, increase customer service scores or rankings, and increase accuracy or timeliness of a given process.

A project selection matrix is used to assist in choosing appropriate projects against pre-established criteria. Table 2.1 demonstrates how potential projects are compared against strategic and business impact and their projected outcome. In the example, projects 3 and 4 are the most likely candidates, based on the pre-established selection criteria.

2.7.4 Determine the Selection Decision Process

Even after the selection criteria are established and the potential projects are compared against the criteria, selection of the final projects can be quite subjective; for example, recent changes in marketplace conditions or internal political influences. The project selection group needs to be clear about how the final decisions will be made. Some executive groups are effective at reaching consensus; others may rely

TABLE 2.1
Sample Project Selection Matrix

Name of Potential Six Sigma Projects	Strategic Impact Rating: 1 = low 2 = moderate 3 = high	Expected Business Impact	Predicted Outcomes
1. Reduce branch teller turnover by 20%	2	Reduce operational costs	$475,000 annual savings in recruiting and training costs
2. Improve cash flow to ensure a minimum of $300K availability of cash on hand	1	Increase ability to respond to market opportunities	$250,000 annual savings in bank fees and interest
3. Reduce customer complaints by 25% on key product	3	Increase effectiveness and efficiency of field support staff	$1,200,000 savings in product support costs
4. Shorten loan processing time by 5 days	3	Increase loan and interest fee revenue	Increased annual revenue of $4,500,000

on the CEO or one or two other powerful individuals to make key decisions. Some groups may even seek to obtain additional buy-in from the functional or business line managers who would be affected by the projects. Regardless of exactly how the project selection decisions are made, it is important to plan ahead regarding how the final selection decisions will be made.

2.8   ASSESSING ORGANIZATIONAL READINESS

From a cultural and process perspective, an organization needs to understand their level of readiness for implementing Six Sigma. An effective and relatively quick way to determine readiness is to use a diagnostic process called force field analysis. This is most effective when it is used in a facilitated discussion among the executives and senior managers. Participants are given small stacks of index cards or Post-It notes. Individually, they identify the factors in the organization that support the Six Sigma implementation and the factors that hinder its implementation. They write down one issue per card and pass them to the facilitator. The facilitator then reviews each card with the group to determine whether it is a factor supporting or hindering implementation. Once all of the factors are segregated, several themes will begin to emerge. From this cultural analysis, the group identifies the strongest themes in support of implementation that can be used internally to promote the benefits of Six Sigma. Then, they identify the major issues or obstacles that will need to be addressed or overcome to have a successful implementation. Figure 2.1 illustrates a force field

DRIVERS	RESTRAINTS
Need to lower costs	Lack of clear priorities
Need to generate revenue	Changing priorities
Adaptability	Overburdened employees
Processes need improvement	Complex projects already under way
Need a disciplined project management approach	Always fighting fires (reactionary management style)
Motivated employees	Lack of buy-in at all levels
Need to improve customer service	Constant short term focus
Process oriented	Rush to fix things
Executive-level support	Resistant to change
Creative employees	Tendency to micromanage
Analytical mindset	Focused on short-term savings
Measurement focused	Incorrect perceptions of real problems
	Limited existing quality infrastructure

FIGURE 2.1 Force field analysis of implementing a Six Sigma initiative within Card One, Inc.

analysis on the driving and restraining forces associated with implementing a Six Sigma initiative at Card One, the company from our case study. (Force field analysis is also used as a tool in the Improve Phase; see Chapter 9.)

2.8.1 HUMAN RESOURCES PRACTICES

Several human resources practices need to be integrated with the Six Sigma implementation to ensure that the organization effectively integrates the Six Sigma methodology into the culture. These practices include establishing and using selection and promotion criteria and developing performance management and reward systems for the individual and teams contributing to the Six Sigma implementation.

The selection criteria need to be integrated into the organization's new hire selection processes as well as internal promotional opportunities. As new candidates are reviewed for their fit with the organization, they also need to be evaluated for their potential in a key Six Sigma role. The candidates selected should embody the skills, experience, and characteristics the organization needs for continued success with the Six Sigma implementation.

Employees selected for the Black Belt or Green Belt roles should see this as an opportunity for job enrichment and career enhancement. Some Black Belts will be selected from the business units or corporate functions for a 12- to 24-month commitment, then return to another role within the business operations. Green Belts usually stay within their business unit while performing the Six Sigma project. When the Black Belts return to the business unit, their value to the organization is as coaches and mentors to the business line employees as well as newly appointed Green Belts and Black Belts. This process provides a way for the organization to continuously transfer knowledge and perpetuate the motivation and commitment among the more experienced Six Sigma players to the novices. Also, in such large-scale transformations as Six Sigma, organizations will need associates from all organizational levels, not just the executives, to promote and sustain the vision.

Employees considered for promotion should exhibit qualities the organization espouses relative to the Six Sigma process. For instance, if a mid-level manager is resistant to the Six Sigma change effort but is promoted for his technical competence, the message to the rest of the organization is that lack of commitment to the Six Sigma effort will not negatively affect professional growth. The real challenge here is not so much with the human resources department or their practices, but rather this manager's superior, who is acting defiantly and allowing his or her department to follow suit.

Performance management measurements also need to be established in organizations that take the Six Sigma implementation seriously. Performance management includes establishing clear performance expectations and developing systems to manage such expectations (e.g., performance appraisals and reviews, goal-setting processes, and reward and recognition programs).

A reward and recognition program needs to be in place very early in the Six Sigma implementation. Examples include team-based bonuses, awards, or social events, as well as individual incentives for outstanding achievement, such as public recognition by the CEO or mention in the company newsletter. Black Belts, Green Belts, and team members should be rewarded early and often, as project successes help Six Sigma gain the momentum it needs for broad acceptance and employee enthusiasm. Human resources managers should be proactive in seeking ways to institutionalize Six Sigma through developing practices to continuously and positively reinforce the initiative. Human resources professionals can also help leaders understand and recognize what new behaviors are required and articulate the connections between the new behaviors and corporate success.

2.8.2 SUSTAINING THE MOMENTUM

How does an organization sustain the momentum of such a broad-scale effort as Six Sigma? There are several ways. As the organization continues to seek out new projects on a regular basis and continues to train Black Belts and Green Belts, new levels of enthusiasm to support the initiative are likely. With a successful Six Sigma track record, leaders should leverage each success to change systems, structures, and policies that no longer fit with the Six Sigma vision. Through such actions, utilizing the Six Sigma methodology becomes an increasingly evident part of the organization's culture.

2.9 COMMON PITFALLS

Organizations place numerous demands on staff members to meet corporate and divisional goals, as well as deal with continuous and overlapping change. Implementing Six Sigma, therefore, requires significant commitment from the organization's leaders and significant effort from the individuals participating on the Six Sigma project teams. As with any change effort, many things can go wrong. Many of these are controllable and with advance planning can be minimized or eliminated so as not to undermine the launch or implementation of Six Sigma. Understanding some of the common pitfalls can help leaders recognize the early signs of problems that can potentially derail an otherwise successful Six Sigma launch.

2.9.1 Lack of Clear Vision

A vision is meant to be an inspirational goal that encourages people to take action toward its achievement and stay committed during the hard times. Thus the leaders of the organization must develop a clear vision and objectives for Six Sigma. This vision should be highly customized to address the organization's own strategic issues; otherwise it will be perceived as another "flavor of the month" initiative. If it is not well articulated, support for the initiative will wane among associates and leaders in light of more pressing issues and demands.

2.9.2 Failure to Adequately Communicate

One of the primary reasons people resist change is that they do not adequately understand the reasons for the change, how it will affect them, and what results are expected. It is the classic, "What's in it for me?" question. This scenario is often created unintentionally, yet in the absence of effective communication, an organization can hinder acceptance of the Six Sigma initiative and negatively affect the progress and outcomes of the Six Sigma projects. When communication has been inadequate in the early stages, insufficient support from the managerial ranks is likely. Over time, this lack of support could result in managers failing to allow their employees to dedicate sufficient time and resources to Six Sigma projects. Managers may simply pay lip service to it. Those who do not like to change in the first place will have a good excuse for dismissing the whole thing as "another failed quality program." Such a pattern of apathy negatively affects the Six Sigma project outcomes and thus becomes a self-fulfilling prophecy. The point is that effective communication practices when launching and sustaining Six Sigma are a critical component to its success.

2.9.3 Lack of Resources

As a company considers Six Sigma implementation, managers typically voice concern about limited resources (employees, dollars, and time) to dedicate to the effort. Their concerns are likely valid. However, the message must be that there is a greater cost to the organization by *not* implementing Six Sigma. This pitfall is overcome by strong executive support in making the necessary commitments upfront to reap the many benefits of Six Sigma down the road. It is truly the CEO's responsibility to ensure that failure is not caused by genuine lack of resources.

2.9.4 Expectation of Quick Results

Characteristic of our Western culture, we want and expect quick results — even when a commitment is made to a long-term process such as Six Sigma. This expectation is difficult to relinquish. Depending on the initial projects selected for

Six Sigma, some short-term gains are likely, namely from those projects that are internally focused and in the form of cost savings due to minimizing waste or effort and increasing efficiency. This is to be expected. Yet, the real gains of Six Sigma come over time as a result of the cumulative effect of numerous projects that contribute to better serving and satisfying customers' needs and improving service or product quality and performance. When improvements in these areas are combined, they translate into a positive impact on the overall value chain provided by an organization to its external customers. The goal is to have early successes with the initial Six Sigma projects, but to also understand that long-term success will come from the results of a combination of many smaller projects that address the longer term issues.

An organization may elect to do Six Sigma pilot projects at first. These projects should be ones that are likely to produce immediate gains to the organization and are not overly complex or ambitious; the idea is to gain quick early successes to build the momentum. After completion of the first round of projects, employees must define the key challenges encountered, to prevent duplicating the mistakes in the next round of more complex projects. Leaders must also assess the need for additional technical training on Six Sigma tools and processes, project management training, or soft-skills training.

2.9.5 PERSONAL AND ORGANIZATIONAL RESISTANCE TO CHANGE

By nature, people tend to fear the unknown and resist changes that are imposed on them vs. changes that they initiate and embrace for their own benefit. On a personal level, individuals can feel threatened and even intimidated by the organization's commitment to adopt the Six Sigma methodology, particularly if they are not clear on the reasons for the change or its intended benefits. Additionally, such a large change can be intimidating because it requires broad-scale participation, even by individuals who may not see the relevance of Six Sigma to their role or function. And, because the Six Sigma methodology involves the use of a fair amount of statistics, many individuals may be concerned about a lack of knowledge in the subject. Employees may experience feelings of job threat and personal insecurity related to their perceived level of competence (or lack thereof), and they may feel a threat of potential loss of political power within the organization, particularly when the Six Sigma team visits their department.

Organizational resistance can be equally daunting, manifesting itself in the form of clashes with the existing culture — the attitudes, values, and practices within the company. The organizations that have fully integrated the Six Sigma approach into their culture have a strong orientation toward continuous process improvement and customer service (to both internal and external customers). In companies that have a strong process orientation, employees are always looking for ways to improve current processes, are open to new ideas from others, and generally have a strong teamwork orientation. Organizations that have not yet developed this type of process-based and customer-focused culture could meet several challenges.

REFERENCES

1. Eckes, G., *Making Six Sigma Last: Managing the Balance between Cultural and Technical Change*, John Wiley & Sons, New York, 2001, p. 207.
2. Pande, P.S., Neuman, R.P., and Cavanagh, R.R., *The Six Sigma Way: How GE, Motorola, and Other Top Companies Are Honing Their Performance*, McGraw-Hill, New York, 2000, p. 7.
3. Pande, P.S., Neuman, R.P., and Cavanagh, R.R., *The Six Sigma Way: How GE, Motorola, and Other Top Companies Are Honing Their Performance*, McGraw-Hill, New York, 2000, p. 9.
4. Honeywell Inc., Annual Report, Honeywell.com, 2000.
5. Slater, R., *The GE Way Fieldbook: Jack Welch's Battle Plan for Corporate Revolution*, McGraw-Hill, New York, 2000, p. 119.
6. Ford Motor Company, Inc., Annual Report, Ford.com., 2000.
7. Honeywell, Inc., Annual Report, Honeywell.com, 1999.
8. Johnson Controls, Inc., Annual Report, JohnsonControls.com, 2000.
9. Dupont, Inc., Annual Report, Dupont.com, 2000.
10. Young, M. and Post, J.E., *Managing to Communicate, Communicating to Manage: How Leading Companies Communicate with Employees*, American Management Association, New York, 1993.
11. Slater, R., *The GE Way Fieldbook: Jack Welch's Battle Plan for Corporate Revolution*, McGraw-Hill, New York, 2000, p. 121.
12. Zenger, J.H., *Not Just for CEOs, Sure-Fire Success Secrets for the Leader in Each of Us*, Richard D. Irwin, A Times Mirror Higher Education Group, Boston, MA, 1996.

3 Work as a Process

3.1 OVERVIEW

Fredrick Taylor, known as the father of scientific management, was a mechanical engineer in the latter half of the 1800s. He proposed that all work is a process and, thus, any work could be studied scientifically and made more efficient through considering the principles of economies of motions. For anyone interested in the study of process improvement, Taylor is a name to remember. Peter Drucker, the renowned organizational and management guru, called Taylor one of the most influential individuals of modern life in the world.[1] Taylor coined the term "management consultant," paving the way for others in the field. (His business card in the early 1900s actually read "Consultant to Management.")

Unfortunately, Taylor's work is remembered largely in a negative context: that of "efficiency expert," for which visions of making people work harder come to mind. Actually, Taylor was considered a friend of the laborer and strived to make life easier for those involved in manual work. One of his most famous studies was how to optimize the process of shoveling coal at Bethlehem Steel Company. He noted that all workers used the same size shovel regardless of the amount being shoveled. A shovelful of rice coal weighed 3.5 lb and a shovelful of iron ore weighed 38 lb; Taylor reasoned that there must be a better way to handle the two materials.[2] "Work smarter, not harder" was the result of Taylor's work at the steel mill; the use of different size shovels appropriate to the shovel loads dramatically increased productivity. In a 3-year period, the work of 500 men was being performed by 140 men.[2] If a Six Sigma team could accomplish that, the savings in today's dollars would be approximately $12.6 million, or $4.2 million per year. Not bad.

The prevailing work environment during Taylor's time was that of the owner-manager, who had very little direct contact with the production worker. Worker management was left to a superintendent who was given total responsibility for the work output and total power over the workers. Superintendents typically viewed the average laborer as lazy, thus requiring direct supervision at all times in order to perform the job. Taylor's pioneering methods in studying human work processes led to the concept of work measurement, in which the content of a job was viewed objectively with definitive inputs of time and outputs of the work product. (The definition of labor productivity is output divided by input, with human work time as the input.) His research led to each job function being allotted a standard amount of time, which provided an objective measure of job performance; no longer were workers subject to the arbitrary tyranny of a superintendent. In addition, because controls were built in through objective work standards, fewer supervisors were required, resulting in overall productivity increases for the factories. For this reason, Taylor must be remembered as having provided a great service to both the industrial workers and the business owners.[2]

A century ahead of Taylor, Adam Smith was remembered more as an economist than an efficiency expert. Smith's observations in a pin-making factory in 1776 led to the idea of specialization of labor, something Henry Ford took to the extreme more than 100 years later.[3] Smith observed that by dividing the task of pin making into four separate and distinct operations, output increased by a factor of five. One person making 1,000 pins per day performing all of the operations could be replaced by 10 workers each performing different operations, resulting in an output of 48,000 pins per day.[2] (Before, ten workers produced 1000 pins per day for 10,000 total pins per day. After, ten workers produced a total of 48,000 pins per day.) One might say that process improvement was officially born in that pin-making factory.

What does all of this have to do with Six Sigma? If you think about Taylor's work at the steel mill and Smith's work in the pin-making factory, they were both applications of studying work processes (transforming inputs into outputs through a series of definable and repeatable tasks) systematically and quantitatively, based on objective analysis of data. Don't ever let anyone tell you that Six Sigma is a new concept.

3.2 VERTICAL FUNCTIONS AND HORIZONTAL PROCESSES

Today's business environments are very different from the steel mills of Taylor's era. However, the concept of work as a process is not. The need for effective management of work processes is the same today as a century ago, only on a much larger scale. Business process management is concerned with managing the entire customer order and fulfillment process, from conception of the product to final payment of the order and extending to service after the sale. Most companies are organized along vertical functions whereas the customer is served horizontally. Innovative companies have moved toward a process-centered, or horizontal, organization structure, in which the focus of the business is on the series of processes that serve the customer. Michael Porter first introduced the concept of a process-oriented company in his 1985 book, *Competitive Advantage: Creating and Sustaining Superior Performance*.[4] Porter focused on processes that enable the value chain. Reengineering guru Michael Hammer took the concept further in his books on reengineering, in which he describes process thinking as cross functional and outcome oriented, both necessary elements for successful reengineering.[5]

The argument for a customer-focused, process-centered organization is a good one. In a typical company, as a customer's request moves horizontally from one department to another, it risks falling through the cracks. Then, when the customer attempts to track down the status of the request, he or she will likely get shuffled again from one department to the next. Rather than directing energy toward effectively fulfilling the customer's request, the departments are busy playing the blame

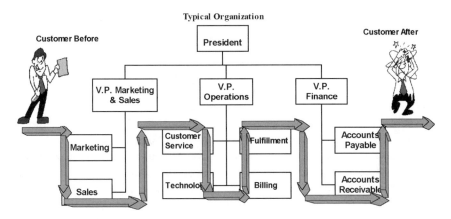

FIGURE 3.1 Process management challenge of managing horizontal processes in a vertically organized company.

game, because of the reward system in place. If the company is typical, the reward system actually encourages the blame game because of conflicting goals and incentives. Sales and marketing are rewarded for bringing in the business, regardless of the terms of the sale, whereas servicing is rewarded for reducing costs. Finance is rewarded for collecting the money and views its customer as the company itself, rather than the purchasing customer who is actually paying the salaries.[6] No one is directly involved with making sure the customer fulfillment and service process is a seamless, defect-free journey through the endless maze of routes along the organizational chart (Figure 3.1).

On the other hand, process-centered companies prevent the customer order from falling through the cracks by instilling a culture of customer-focused service centered around horizontally managed processes. These companies maintain an outward-focused customer perspective, rather than the typical inward-focused viewpoint. Most employees view the process as a series of linked actions and understand how their process fits into the larger picture of the customer order, fulfillment, and service cycle. Process performance measures are emphasized and are routinely reported and discussed in relationship to the impact on the customer.[7]

As previously mentioned, Japanese industrialists came to the realization long before their American counterparts that the quality of the process dictates the quality of the product. U.S. companies have not succeeded as have the Japanese in effective process management, and much research and effort has gone into figuring out why this is the case. Cultural differences are part of the answer. To build and maintain a quality process, a company must remove the barriers that cause long cycle times and poor quality; quite often, the vertically organized command-and-control structure of Westernized companies is the main barrier. Along with actively managing their key processes, Japanese companies have historically focused on decreasing cycle times at all process levels, from the strategic to the tactical. They realize that long cycle times are at the root of many poor quality issues. On the other hand, U.S. businesses tend to react to poor quality by adding quality inspections into the process, which adds more time. Their logic is that the customer must receive a high-quality product, no matter what it takes. Often the poor quality is a temporary situation resulting from a special cause, and the extra inspections "temporarily" added become ingrained in the process. Not that anyone is intentionally looking to lengthen the cycle time of the process — at the time it seems perfectly logical.

Research shows that long cycle times are a major dissatisfier among customers; in other words, there is no credit for achieving cycle time goals, but there are many negative consequences for not achieving them.[8] Increasing the time from request to fulfillment causes a domino effect in the business. Not only is there the major issue of customer dissatisfaction with each passing moment, there are more opportunities for things to go wrong and there is greater work in process, which means more capital tied up in the business and not earning a return.

In managing processes, the individuals in a position to take corrective action on the process are usually the ones who do not see the issues. And the employees who work on the front line and deal with the problems day in and day out are not in a position to fix them. It is important that the business leaders, the ones who can allocate the resources to take actions, understand the fundamentals of process management. This is especially the case in a transactional services environment, where about 85% of the business capital is human capital. Business leaders need to understand that most of the business processes are human processes and, thus, invest in hiring, training, and managing this high-cost, high-return asset. They need to understand the intricacies of the customer order and fulfillment process, with the knowledge that the customer gets what the process is designed to give. And they need a deep appreciation of the fundamentals of process management: understand the customer's requirements and design the business processes to meet those requirements. It sounds deceptively simple and yet it remains an elusive goal for most businesses.

REFERENCES

1. Drucker, P.F., *Management Challenges for the 21st Century*, HarperCollins, New York, 1999.
2. Hicks, P.E., *Introduction to Industrial Engineering and Management Science*, McGraw-Hill, New York, 1977.
3. http://www.bhdtech.com/engineer_in_the_enterprise.htm, 2001.
4. Porter, M., *Competitive Advantage: Creating and Sustaining Superior Performance*, Free Press, New York, 1985.
5. Hammer, M. and Champy, J., *Re-Engineering the Corporation: A Manifesto for Business Revolution*, Harper Business, New York, 1993.
6. Shapiro, B. et al., Staple yourself to an order, *Harvard Business Review*, July 1992, pp. 113–122.
7. Petrozzo, D.P., and Stepper, J.C., *Successful Reengineering: Now You Know What It Is — Here's How To Do It*, Van Nostrand Reinhold, New York, 1994.
8. Evenson, A., Harker, P.T., and Frei, F.X., *Effective Call Center Management: Evidence from Financial Services*, The Wharton School, University of Pennsylvania, Philadelphia, 1999.

4 Voice of the Customer

4.1 IMPORTANCE

A customer-focused organization is not only a business imperative for success in the 21st century, it is also an integral part of the Six Sigma implementation. Management guru Peter Drucker writes in his recent book, *Management Challenges for the 21st Century,*[1] that organizations need to place a higher emphasis on what customers value than simply the product or service that the organization has to offer. The reason is that customers' interpretation of "value" invariably differs from the company's perception of what customers value. Accepting this paradox is the starting place for organizations to become customer focused.

Many companies claim they are customer focused, yet show little concrete evidence to substantiate it. An organization that is truly committed to a customer-focused strategy has multiple customer feedback channels and a structured methodology for integrating the data into its product development and customer service delivery processes. These organizations are more likely to truly understand and meet their customers' needs.

What is the real benefit to an organization for being customer focused? Further, how does a company put the systems and practices in place to get beyond the advertising slogan of being customer oriented? Regardless of whether a company is manufacturing a product or providing a service, marketplace competition is always present. The competition forces companies to continuously refine how their product or service differs from its competitors. In the absence of such differentiation, customers are likely to view the product or service options available as commodities and, thus, choose the one that fits their budget. Gaining or sustaining a market leader position requires structured and intelligent systems that continuously gather requirements from current, lost, and potential customers. The market intelligence is then actively integrated into the nervous system of the company, including product development, marketing and sales, operations, and distribution. The process of gathering, analyzing, and integrating customer input back into the organization's operations is called the "voice of the customer" (VOC) and is one of the most critical components of Six Sigma.

Six Sigma provides a structured approach to improve and develop customer-focused processes. It begins by linking the rationale for becoming more customer focused to the business strategy. The leaders of the company set the tone for whether "customer focused" is an advertising campaign slogan or a way of doing business. Clearly defining and advocating the whys and wherefores of being customer focused should be at the top of the priority list for companies who expect to reap the benefits of Six Sigma. Among these whys and wherefores are:

- Why is focusing on meeting customers' needs a business imperative?
- How will the business expect to benefit from such an effort?
- What internal changes will need to occur?
- What is the cost of NOT being customer focused?

The answer to the last bullet point is quite simple. Research indicates that it costs 10 times more to gain a new customer than to hold on to an existing one.[2] A quick way to calculate the financial impact of losing a customer is to identify which customers have "fired" the company within the past year. Next, multiply each of those customers' annual revenue by 5, 10, or even 20, for the number of years they could have continued to contribute to the company's revenue stream. Then, combine those totals to arrive at the total estimated lost revenue of those customers. That simple process will capture the attention of even the most stubborn skeptic. Clearly, there is an enormous financial reward for running a customer-focused organization.

Moving to a customer-centric philosophy requires fundamental actions that are deceptively simple on the surface. However, it if really were that simple, I would not be writing this book and you would not be reading it. The fundamental steps in becoming customer focused are:

1. Identify the customers, both internal and external.
2. Develop VOC systems for continuously collecting and analyzing customer data, using multiple channels from various customer segments.
3. Translate the VOC into product and service requirements, or CTQ characteristics.
4. Incorporate the VOC data into the organization's nervous system, where day-to-day decisions on allocation of resources, new markets, new products and services, new processes, and new technologies are made from the perspective of meeting and exceeding customer CTQs.

4.2 IDENTIFY THE CUSTOMER

Customers come in two flavors: external and internal. External customers are those who either purchase or are the final recipient of the service or product. They are also regulatory or government agencies as well as shareholders, all of whom have specific requirements. To maximize the return on external VOC dollars, a market segmentation strategy is suggested. Effective segmentation occurs on multiple levels, such as distinguishing by types of service or product, and then by their impact on the organization's bottom line. In absence of customer segmentation, the company is likely using a "one size fits all" approach to attempting to meet customers' needs — an inefficient and ineffective way to capture VOC.

Multiple perspectives of the company's customers, including current, former, and prospective customers, is required. Some organizations adopt an arrogant view toward former customers and disregard the reasons they switched to the competition. Such an attitude can perpetuate a negative effect on the business, especially if customers are defecting for reasons that are not fully understood. Also, many organizations take their existing customer base for granted and often fail to understand their evolving requirements. Think about it — what percentage of your customers have static requirements?

Organizations seriously interested in collecting and analyzing customer data by listening to the VOC are less likely to encounter the pitfalls of losing customers and failing to build customer loyalty. These customer-focused companies have the necessary closed-loop feedback systems to quickly understand the reasons customers defect and take prompt action to prevent more of the same.

Internal customers are individuals and groups that rely on another department's output to carry out their work. For example, a group of underwriters in a bank prepare information for the loan officers (internal customers), who in turn provide the finished product to the external customer. Each group that contributes to the external customer product is expected to perform their function accurately and timely, in order for the next group to meet their customer requirements. There are many levels of internal customers within an organization, ranging from business managers who rely on accurate and timely financial reports, to the front-line customer service agent who relies on the information technology department to provide adequate uptime of the servicing system.

4.3 COLLECT VOC DATA

The methods for capturing VOC range from monitoring inbound phone calls to formal surveys and focus groups. The company should have ongoing processes for continuously gathering customer data. This facilitates the Six Sigma team efforts in readily understanding the customer CTQs for their project. Several VOC data collection methods are discussed in this section.

4.3.1 SURVEYS

When designed and applied properly, surveys provide powerful insight into the customer's perspective. Specifically, survey data can define and prioritize the key elements that contribute to customer satisfaction. Two common types of VOC surveys are relationship and transactional surveys. Relationship surveys ask questions relating to the entire relationship with the company whereas transactional surveys ask questions relating to a specific transaction, such as a billing statement or a recent call into customer service. Relationship surveys are designed to understand the customer's needs relative to the entire relationship with the company, such as overall satisfaction with the entire line of services or products offered. These surveys are longer and sent less frequently than transactional surveys. Transactional surveys are sent immediately after a specific transaction has occurred between the company and the customer and are used to understand process issues relating to the transaction. New product development and marketing departments typically send relationship surveys whereas service operations and quality departments are interested in the

---

Card One Customer Survey

Please rate your level of service satisfaction by using the following rating scale in response to the questions below:

  1= Excellent   2 = Very Good   3 = Satisfactory   4 = Fair   5 = Poor

1.	Overall, how would you rate the quality of service provided by Card One?	1	2	3	4	5
2.	How would you rate the response time for your call to be first answered?	1	2	3	4	5
3.	If you were placed on hold, how would you rate the waiting time?	1	2	3	4	5
4.	How would you rate the time it took to speak with the person who finally assisted you?	1	2	3	4	5
5.	What level of knowledge about your question did the representative display?	1	2	3	4	5
6.	How accurate was the information that the representative provided to you?	1	2	3	4	5
7.	How would you rate the extent to which your issue was resolved?	1	2	3	4	5
8.	How satisfied were you with the level of courtesy that the representative displayed?	1	2	3	4	5
9.	To what extent did you feel that the representative treated you as a valued customer?	1	2	3	4	5
10.	How would you rate the quality of this call compared to your last call made to Card One?	1	2	3	4	5

Additional Comments:

Thank you for your time and valuable input! Please call us at 1-800-CARD-ONE for your comments and questions.

FIGURE 4.1 Card One transactional customer satisfaction survey.

results from transactional surveys. Regardless of the type of survey, there are benefits to both, including:

- Structured and consistent data collection method
- Ability to gather specific information on customers' wants and needs
- Ability to target surveys to specific customer groups
- Ability to segment and prioritize customer groups
- Relatively efficient and cost effective

The drawbacks of surveys include low response rates and limited ability to obtain open-ended responses. Response bias, which is the tendency of the customer to respond only when extremely irate or extremely pleased, is also an issue, especially in mailed surveys. This is sometimes called the barbell effect, because response data exist on either end of the spectrum instead of toward the neutral center.

Figure 4.1 provides an example of a transaction-based VOC survey for our case study, which is to reduce call volume in a large inbound call center for Card One, a credit card and financial services company.

4.3.2 INTERVIEWS AND FOCUS GROUPS

The ability to meet face to face with current or potential customers provides enormous benefit to an organization. Interviews conducted on a one-on-one basis via phone or in person offer obvious advantages over a paper-and-pencil survey instrument. The opportunity for the interviewer to ask open-ended questions to understand the customer's perspective and the ability to delve more deeply into the customer's responses are two of those advantages.

Interactive surveys can be relational or transactional in nature. Transactional surveying soon after a specific business transaction is a powerful way to capture VOC, because the customer's memory of the transaction is fresh. Six Sigma teams can use this type of customer feedback to determine the quality and satisfaction levels for a specific process.

Focus groups generally consist of a meeting with approximately 6 to 8 people plus a group facilitator. In this format, the facilitator can uncover feelings, attitudes, and biases that either motivate a customer to utilize the product or service or discourage him from doing so. The discussion format encourages open exchange of ideas and issues that are may not occur in one-on-one surveys and interviews. Benefits of interviews and focus groups include the following:

- Ability to probe for in-depth understanding of customer's responses and requirements
- Effective for targeting specific needs or the needs of specific customer segments
- Potential to increase good will with targeted customer groups by showing an interest in understanding their particular wants and needs

The drawbacks of these methods include the following:

- Time-consuming to conduct
- Difficult to consolidate and codify qualitative responses
- Variability of data quality and volume with multiple interviewers and focus group facilitators
- Small size of focus group and one-on-one interviews limits the number of customers that an organization can effectively reach
- More costly than surveys

4.3.3 DIRECT CUSTOMER OBSERVATION

Organizations that are serious about their VOC efforts find creative opportunities to understand their customers' experiences with their services and study it as closely as possible. Visiting a customer's site to observe, analyze, and understand how the customer uses the company's product or services in one method of in-depth VOC analysis. Data gathered this way is invaluable in understanding who the various users are within the customer organization, what their particular needs are, and exactly how they are using the product or service. This type of meeting is most effective when it is an exploratory visit to identify customer behavior, issues, and concerns relative to the product or service rather than an opportunity for a disguised sales call. The added benefit for the provider organization is to build good will in the customer–provider relationship by taking a special interest in the customer.

4.3.4 Act As Customer for the Day

Many organizations have no idea what their customers experience until members of the company take themselves through the same processes their customers would in using their product or service. For example, take a simple transaction like phoning your customer service hotline: How long is it before the call is answered? How does the recording or person answering the phone sound? Are you put on hold? If so, for how long? What is the response time before getting directly to someone who can help you? Do you feel that the representative treated you as a valued customer? Notice that most of these questions relate to the *process* and not necessarily the problem that triggered the call. A recent research project showed that customers are three times more likely than the service provider to recall the "human element" of a transaction; the service provider tends to focus on the "business element" of the interaction.[3] Another study at a financial services firm's call center determined three key factors influencing the quality of its service delivery. These critical elements had to work together to provide overall quality service to their customers: effective people, effective processes, and effective information technology.[4] Acting as customer for the day is usually an eye opener, unless the company has a well-indoctrinated and near-perfect customer service culture.

4.3.5 Customer Complaint Processes

A low-cost and effective method of capturing VOC is to analyze the disputes and complaints received through the company's customer service departments. Customer e-mails and quality assurance monitors can provide a wealth of data on this subject. Properly organized, the data are a rich source of analysis to understand systemic issues in the service processes and policies of the company.

4.4 CRITICAL-TO-QUALITY (CTQ) CUSTOMER REQUIREMENTS

The goal of VOC is to understand specific customer requirements and acceptable ranges of performance as defined by the customer. Continuous feedback of customer data, through methods discussed in previous sections, offers distinct advantages to a company getting started in Six Sigma. Not only does this mean that the Six Sigma team has readily available VOC data to get started, it also indicates the seriousness of the company in listening to their customers.

Voice of the Customer (VOC)	VOC Theme Statement	Critical to Quality (CTQ) Attribute
The service here is lousy! There are no salespeople in sight!	Customer needs readily available sales staff.	A sales person will greet customers within 60 seconds upon their entering the department.
The application approval process takes too long. I don't know when and if my loan will ever get approved.	Customer needs a quicker response time after submitting their loan application.	Customers will be notified of their application results within 3 working days after submittal.
There's no support for how to use this equipment! The instructions are confusing and I can't get it to work for me.	Customer needs company assistance in learning how to use the equipment.	New customers will receive hands-on training within 48 hours of equipment installation.

FIGURE 4.2 Translating VOC into CTQ attributes.

VOC data is necessary to determine the gap between the customer's CTQ requirements and the current levels of the process. Once the size of the gap is identified and quantified, the Six Sigma team can begin to focus on analysis and improvement of the process.

4.4.1 CTQ Defined

A customer's CTQ characteristics are defined as "product or service characteristics that must be met to satisfy a customer's specification or requirement."[5] In a Six Sigma approach, CTQs are gathered through various methods and initially kept in the customers' verbatim form. Keeping the phrases in the customers' words minimizes the chance of the Six Sigma team projecting their own interpretation of the customers' needs, introducing bias into the research. Next, the customers' needs are grouped and analyzed into related theme statements. Then, the customer requirements are translated into terminology that reflects the customers' CTQ characteristics and are stated using measurable terms (Figure 4.2).

The process of identifying the customers' CTQ characteristics is accomplished through group discussion, with input from team members and other employees who are also familiar with the customers' needs. As Figure 4.1 illustrates, understanding the context of the customers' comments is required to effectively develop the theme statements and CTQs. In other words, the process of translating the VOC into CTQ characteristics cannot occur within a vacuum; it requires input from sources close to the customer. Each CTQ should also be further defined in terms of ranges of customer acceptability. For example, a customer may be *highly satisfied* if greeted by a sales associate within the first minute, *satisfied* if greeted within 5 min, and *highly dissatisfied* if acknowledged only after 10 min. The various ranges within a CTQ measure and the corresponding levels of customer satisfaction are also very important for the project team to understand. Once the CTQs are identified, they should be validated with a representative sample of the organization's customers. This validation process is an integral part of confirming (and modifying if necessary) the customers' requirements.

4.4.2 Business CTQs

The concept of customer CTQs is well known throughout the Six Sigma field. However, business CTQs are not as well defined. Business CTQs are those critical-to-quality measures that must be attained for the business to survive. Return on investment and profitability measures are examples of business CTQs. Internal efficiency measures that the customer may not care about directly are often business CTQs, such as average handling time of calls or the percentage of phone calls that are handled entirely by the interactive voice response system.

4.4.3 Integrate CTQ Measurements into Reporting Systems

After the Six Sigma project is completed, any customer and business CTQs that are not already included in management reporting should be added. The philosophy is that if the measure is critical enough to be part of a Six Sigma project, it is most likely critical enough to be tracked on an ongoing basis.

4.5 FINAL NOTE

Remember that customers' CTQs are not static; change will occur whether you are prepared for it or not. The process of continuously analyzing customers' needs and matching against the internal process capabilities takes time, effort, and patience. However, the benefits to the organization are bottom line profits through knowing their customers better than the competition. The chapters on the Six Sigma DMAIC approach provide a solid foundation for making sure the internal processes are capable of meeting the identified CTQs profitably.

REFERENCES

1. Drucker, P.F., *Management Challenges for the 21st Century*, HarperCollins, New York, 1999.
2. Gitomer, J., *Customer Satisfaction Is Worthless, Customer Loyalty Is Priceless*, Bard Press, Austin, TX, 1998, p. 60.
3. Berrey, C., et al., *When Caring Is Not Enough*, Achieve Global, Inc., Tampa, FL, 2000.
4. Evenson, A., Harker, P.T., and Frei, F.X., *Effective Call Center Management: Evidence from Financial Services*, The Wharton School, University of Pennsylvania, Philadelphia, 1999.
5. Simon, K., "Customer CTQs, Defining Defect, Unit and Opportunity," sixsigma.com; Naumann, E., "Capturing the Voice of the Customer," via ASQ Six Sigma forum — articles, asq.com.

5 Project Management

5.1 PROJECT MANAGEMENT CHALLENGES

Managing projects may be one of the most difficult and challenging assignments of a professional person's career and managing a Six Sigma project is no different. The project manager is ultimately responsible for the success of the project, yet has no formal control or authority over the project team members who are involved in performing a significant amount of the work. Many decades ago, this model would not work because the project manager would need to have ultimate control. But, in today's team-oriented business environment, effective project managers know their success is determined by their ability to use effective influence to motivate and keep team members productive, without using force or intimidation. It is important for project managers, particularly those early in their careers, to recognize that it is not the project's task schedule or control plan that gets the project completed, it is the people.

For the most part, people in organizations want to work and contribute in meaningful ways. The relevance of their project provides meaning, but the quality of human interaction within the project teams can also strongly influence their ongoing commitment and motivation. Project managers can maximize their effectiveness by first understanding aspects of the company culture and then applying the practices of project management that best fit that culture.

Fortunately, the phased DMAIC approach lends itself to structured project management. The project phases are known in advance and thus the project team has a "roadmap" for reference from the beginning. It is my experience that each Six Sigma project progresses differently — some have a difficult time moving beyond the Measure Phase whereas others get bogged down in the Improve Phase. In almost all projects, however, the first two to three phases seem somewhat chaotic, with many different paths and directions the project can take. The good news is that at some point the project seems to coalesce, with clear direction on what the true issues are and how to best approach solutions.

5.2 PROJECT CULTURE

A "project culture"[1] is one in which an organization's employees are accustomed to working with others in cross-functional, temporary teams. Team members from various departments and positions within the organization come together on an ad hoc basis to accomplish a specific task or goal. After the team has accomplished the goal, it is disbanded and the members return to their full-time jobs; this is how most Six Sigma projects are conducted.

5.2.1 STRONG PROJECT CULTURES

Team members in a strong project culture develop a familiarity and skill in working with others outside of their formal and immediate work groups. They have a stronger sense of their role as a team member in contributing to the team's purpose and see that role as equally important as their primary job roles. High employee empowerment tends to correspond with these cultures, whereby employees who are close to the work processes are involved in continuous improvement projects and in contributing to decisions that affect their work. Team members in a strong project culture are also more likely to be cooperative vs. political or territorial, compared to a weak project culture.

The primary role of the project manager is to use influence skills to get others to accomplish the project work, even though they lack formal authority and control. Parallel to this is responsibility for making sure the project stays on track and is delivered on time. Effective project managers have a strong task and goal orientation in order to keep the project on schedule, but are able to balance those demands with their effective interpersonal skills, especially their ability to motivate and obtain results from others. A project manager who disregards the interpersonal aspects of managing projects through managing others is often viewed as a taskmaster. This management style will likely alienate the team members, resulting in lower morale and motivation and compromised project outcomes. Likewise, project managers who display a greater concern for the team members' well-being than for the task at hand could find themselves falling quickly behind on the project schedule. A balance between people and technical skills is required. The organizational culture usually determines on which side of the scale the project manager finds success. In a highly structured, hierarchical culture, the balance is toward the technical side, whereas a more open, flexible culture requires greater people skills. Often, when a project manager is less than successful, it is a mismatch between company culture and personal style, rather than a skill or will issue.

Effective project managers realize these distinctions and consciously balance the tasks and people equation. They are effective at giving team members a sense of empowerment by giving them clear responsibilities as well as clearly defined goals and tasks and they hold team members accountable. They also know how to keep the team members motivated, by providing positive reinforcement for accomplishments and conveying encouragement and confidence in the team's ability to complete future tasks.

Common sense dictates that a strong project culture increases the likelihood for success of a Six Sigma implementation.

5.2.2 WEAK PROJECT CULTURES

A weak project culture is reflected in an organization that has little or no cross-functional project teams. This is not a criticism; however, it is important to understand the difficulties Six Sigma teams will face when a project-oriented initiative is introduced. The Black Belts and other employees who are responsible for managing Six Sigma projects and participating on project teams will need to evolve into their roles, learning the skills and behaviors that contribute to team success.

Greater emphasis on formal training in project management as well as team and communications skills is required in this type of company. I have (unfortunately) observed many cases in which a Six Sigma implementation effort was more painful than it ought to be because of the lack of project management and team facilitation skills. It is a mistake for companies to just send some of their best and brightest employees to a 4-week Black Belt course and expect success. These individuals need to master significant new technical skills, but just as important are the skills required to successfully control project resources and accomplish work through teams.

5.3 PROJECT MANAGEMENT PROCESSES

In the context of Six Sigma, projects are undertaken to achieve specific business outcomes within a specified time utilizing the DMAIC methodology and principles from the project management discipline. (The DMAIC model is quite detailed and is covered in subsequent chapters as part of the project's execution phase.) Below are several of the critical steps to successfully manage projects.

5.3.1 BUILD THE TEAM

A typical Six Sigma team includes the Black Belt or Green Belt as project manager and team members from a wide variety of backgrounds and knowledge levels. It is not uncommon for the project Champion to be on the team. Team members are usually identified during the Define Phase of the project, when the project charter is developed (see Chapter 6). Project team members are selected based on their background, experience, and skills that they can contribute to the team. A mistake many companies make is to select team members from employees who have free time on their hands. Frequently, they have free time because someone has been reluctant to assign responsibility to them — maybe for good reason. Be mindful of this in selecting the Six Sigma team members.

Another common mistake is not involving the employees who will be responsible for implementing the changes. Although this seems elementary on the surface, it is sometimes difficult to understand in the initial stages of the project where the implementation responsibility will fall. This happens because the path of Six Sigma is one of discovery and the data will lead the team to where the improvements which may involve organizational changes are needed the most.

When selecting participants for the project team, special consideration needs to be given to the project itself. If it is a highly technical project, it makes sense to include highly technical employees. If it is a project in the customer service area, then certainly customer service employees need to be represented. However, some general guidelines should be followed for any type of project. Involve individuals who have the interpersonal skills to work in a team environment, are open to new ideas, and who have a minimal level of analytical skills. Section 5.4 provides an overview of a personality typing tool (the Myers–Briggs™ inventory) that is specifically geared toward facilitating teamwork. The tool helps team members understand their own learning and decision-making styles as well as those of their fellow team members. This promotes teamwork by viewing other team members' personality styles in an objective manner. So when fellow team member Roy refuses to let go of an issue after the team has already spent 5 hours on it, at least you know that he is not intentionally being difficult — it is just that he has to understand all facets of an issue before moving on to the next.

The Black Belt and Champion need to play critical roles in identifying team members because they are ultimately responsible for the team's success. They also need to identify key stakeholders and how they will be involved and kept informed of the team progress.

5.3.2 PLAN THE PROJECT

Using the DMAIC framework, the Six Sigma project manager needs to conduct the initial project planning. Most likely it is the Black Belt's role to manage the project, although this varies from company to company. Initial project planning involves identifying the major project tasks, estimating the time involved, and identifying who is responsible for each task within each of the five project phases.

Once the team is selected, the project manager needs to establish a leadership role through initiating meetings and team communications. The Six Sigma team should plan to meet on a routine scheduled basis. The suggested minimum is once a week, although more frequent meetings are desirable, especially at the beginning of the project. If the company uses an electronic calendar system, the meetings should be set up as recurring on all the team members' calendars. One of the first tasks of the Six Sigma team is to develop the project charter; a communication plan is a key element of the charter and is discussed further in Chapter 6.

5.3.3 CONTROL THE PROJECT

The project plan is the living, breathing document that provides the overall direction for the Six Sigma team. Fortunately, the structured DMAIC approach provides an ideal framework for project planning. If the project is properly scoped during the Define Phase, the boundaries for the project plan should be clear. However, it is extremely easy for a Six Sigma project to quickly spiral out of control. This is especially true during the Measure Phase when a variety of data is being gathered from many different sources and the team has not had time to analyze and understand the data. The Black Belt and other team members must recognize the tendency to overcollect data during the early phases of the project, while still trying to establish a framework for the ultimate direction it will take.

Another area where there is a tendency for the project to lose control is in the Improve Phase. Because developing alternative solutions is a primary goal of this phase, team members can easily go in many directions seeking an optimal solution. Although the Six Sigma toolset is designed to provide a structured approach toward problem solving, the Improve Phase by its very nature is less structured than the others.

Some Six Sigma projects lend themselves to incremental improvements whereas others seem to take on a life of their own, scope or no scope. In my experience, the Six Sigma projects with the greatest opportunity for losing control are those with a significant amount of data to collect and analyze. The projects that have the least amount of data usually go the fastest. The Six Sigma team needs to be aware of the many points where a project can quickly lose control and take conscious steps to avoid this situation.

5.3.4 CONDUCT TOLLGATES

Project tollgate meetings provide an opportunity for the Six Sigma project team to update the advisory board members on the status of the project. Many companies require a tollgate meeting at the end of each of the DMAIC Phases; others require only one or two meetings with the executive team during the course of the project. Depending on the Six Sigma team structure in place, the Black Belt, Green Belt, Champion, or team members are responsible for presenting the project results.

Tollgates need to be held early in the project to ensure the team is on the right track; the opportunity for interactive discussion of the project progress (or lack thereof) with those responsible for allocating resources is a key element of success. However, it is also important to maintain the objectivity that is required of Six Sigma — the potential danger in tollgate meetings is for a few influential executives to lead the team down a politically appealing path, without full consideration of the data.

The most important tollgate meeting is at the end of the Improve Phase. This is typically when the project recommendations are approved and implementation planning begins. All of the hard work of the Six Sigma team is evident at this meeting, and a properly conducted project will leave little room for disagreeing with the fact-driven recommendations. Think of it as an attorney at trial who leads the jury down the path of justice as the right thing to do. Sound improvement recommendations supported by facts and thorough analysis are the outcome of a successful Six Sigma Improve Phase tollgate. (If only it were that easy!)

5.3.5 CLOSE THE PROJECT

The definition of the "end" of the Six Sigma project varies from company to company and from project to project. Many Six Sigma teams disband at the end of the Improve Phase, handing over the implementation to the business line managers. Others assist in implementation and detailed monitoring of the results. Regardless of the formal ending of the project, there needs to be a formal sign-off process within the DMAIC project model. As with the tollgates, the particular Six Sigma implementation model the company chooses will drive the sign-off process.

Some companies conduct a "postmortem" on the project, in which the team and other stakeholders perform a quick analysis of the lessons learned from the project. These could be the top challenges the team needed to overcome with respect to managing people or processes. The teams document their learnings and the strategies used for overcoming challenges. It is a good idea to keep track of the lessons learned from all the Six Sigma teams in order to analyze the data for patterns. The same issues are often repeated throughout the DMAIC projects — ongoing tracking allows retrospective insight into the project challenges, and plans can be made to prevent them in the future.

Visibly celebrating the success of the project is important. Even small events to show appreciation to the team members are better than letting the milestone pass with no recognition of their hard work and dedication. Special lunches, framed certificates, plaques, or gifts that have significance to the project are suggested when formally disbanding the team. The opportunity for appreciation and recognition from the company's executive staff is highly motivating to other employees who may be wondering if they should get involved with a Six Sigma team.

5.4 TEAM TYPING

The Myers–Briggs Type Indicator® (MBTI®) is an assessment tool that helps people understand their own and others' preferences and temperaments. It uses four continua with opposite factors to develop a composite profile of a person's natural tendencies.[2]

The four continua are as follows:

Source of Energy

Extroversion (E)   or   Introversion (I)

Information Gathering

Sensing (S)   or   Intuition (N)

Decision Making

Thinking (T)   or   Feeling (F)

Lifestyle

Judging (J)   or   Perceiving (P)

The results of the MBTI questionnaire identify an individual's overall preferences based on the strength of those preferences within the four continua. A total of 16 different preference combinations, or profiles, exists. The MBTI assists in analyzing the strengths that each member brings to the team; the job of the project manager is to link natural preferences with tasks and processes. The more similarity there is among team members, the quicker they understand one another. Team members with differing profile types may take more time to understand each other and may not always see things the same way. A good example is to ask an "intuitive" type what time it is; he or she might say, "it's around 4:00." But a "sensing" type (one who deals in facts and figures) might say "it's 3:48 p.m.," and so it goes. It is actually healthy for a team to have team members with differing profiles so that numerous perspectives can contribute effectively to problem solving and task execution.

5.5 TEAM STAGES — UNDERSTANDING TEAM DYNAMICS

As a Six Sigma team is formed, do not expect them to immediately propel into effective action. All teams go through predictable stages of team development as the team members get acquainted and understand their respective work styles. Below are the four stages of team development, as described in *The Team Handbook*.[3]

5.5.1 FORMING

The initial stage, "forming," is when team members get together for the first several times. At this stage, the team members are generally in a more observant mode than normal, gathering information from what they see, hear, and learn from their colleagues. During this process, they are assessing one another's roles and relationships to each other. They are also, individually and collectively, clarifying their understanding of their common goal and the logistical aspects of their team formation (e.g., ground rules, meeting frequency, etc.).

5.5.2 STORMING

During the second stage, "storming," it is common for the team members to drop the formality with each other that was present during the previous stage and now exercise some of their individuality. This will manifest as the team members assert their power and control with respect to their own ideas and opinions of the project details. During this stage, the team must learn how to balance the power forces within the team in order to move on.

5.5.3 NORMING

The next stage, called "norming," is when the team becomes cohesive. They have survived the storming stage and are learning how to cooperate with each other as the behavioral norms take root. This is an exciting stage because the team begins to gain its own identity and the team members develop an esprit de corps. During this stage, the team members are open to sharing ideas and differing opinions with each other as they exchange their focus on personal agendas for that of the project goals.

5.5.4 PERFORMING

By the last stage, "performing," the team has matured and is operating at a high level of capacity. Great camaraderie and mutual support exists. This stage is also characterized by a powerful team identity and high morale. The team can engage in effective problem solving together, based on the trust developed through open and productive dialogue, because the team members have coalesced and are directing their efforts toward accomplishing the Six Sigma team's objectives.

5.6 CHARACTERISTICS OF EFFECTIVE TEAMS

5.6.1 COMMON MISSION

The first and fundamental aspect of an effective team is one whose members share a common mission or purpose. Fortunately for the Six Sigma teams, the Define Phase focuses on clarifying the purpose of the project through developing a project charter. Because the charter lays out a clear foundation and direction for the entire project, there is likely to be a higher level of mutual commitment to see the project through to completion. In ineffective teams, team members may subjugate the team's goals for their own personal gains.

5.6.2 OPENNESS AND TRUST

Openness and trust among team members is evidence of an effective team. Without a climate of openness and trust the team cannot effectively discuss various issues that arise during the Six Sigma project. The corollary is a team with one, two, or more dominating or politically motivated team members who squelch the inputs of others thereby diminishing the team's strength and effectiveness. The Black Belt or Champion is responsible for making sure this does not happen.

5.6.3 COMPLEMENTARY SKILLS

Teams whose members have complementary skills and personalities achieve greater success. With a diverse set of skills and perspectives, the team can share responsibilities for the various tasks depending on their respective skill levels and natural preferences. Teams that are too homogeneous are likely to either clash too frequently or not conflict at all. Both scenarios are problematic and limit the quality of the project outcomes.

5.6.4 EFFECTIVE DECISION MAKING

Team members need to utilize various decision-making processes. These include understanding the different ways of evaluating issues prior to making decisions, through use of decision trees and weighting factors, as well as the pros and cons of voting and the skills in reaching consensus. Effective teams are able to reach consensus on most issues. The benefit of consensus is that the team is more likely to have strong buy-in to the decision and outcome. This ability to reach consensus actually occurs as a result of the team facilitating open and effective dialogue.

5.7 SUMMARY

Managing projects is a complex process that requires interpersonal finesse, formal project management experience, and, as it relates to Six Sigma, understanding the DMAIC methodology. The failure to recognize the need for effective project management is a primary reason Six Sigma teams do not achieve optimal results. To provide the greatest opportunity for success in implementing such a significant change effort as Six Sigma, it is wise to allocate resources in the early stages to training and awareness in both project management and team facilitation skills. Do not let your company be a victim of the penny-wise and pound-foolish mentality that is surprisingly common in many companies.

REFERENCES

1. Graham, R.J., *Project Management As If People Mattered*, Primavera Press, Bala Cynwyd, PA, 1989.
2. Hirsh, S.K., *Introduction to Type and Teams*, Consulting Psychologists Press, Palo Alto, CA, 1992.
3. Scholtes, P.R., *The Team Handbook: How to Use Teams to Improve Quality*, Joiner Associates, Inc., Madison, WI, 1998.

6 Define Phase

6.1 INTRODUCTION TO THE DMAIC PHASES

Research indicates that for a process with high defect rates, human inspection can detect defects with 80% accuracy. At lower defects rates, inspection is 90 to 95% accurate.[1] However, as the sigma level approaches six, it is absurd to think that human inspection can detect 3.4 defects out of 1,000,000 inspections — not to mention the inefficiency of such an inspection process.

Recall from our earlier discussions on process quality vs. product quality that Japanese manufacturers have been able to consistently produce high-quality products that require little inspection before shipping. They do this by focusing on the quality of the *process* used to manufacture the product. In most situations, a defect-free process produces a defect-free product.

Together, these two main points — (1) quality cannot be inspected into a process and (2) a quality process is required for a quality outcome — make a compelling case for applying a sound and proven approach to designing error-free processes for businesses. This chapter and the four that follow describe a method for achieving these objectives: DMAIC. Any company considering implementation of Six Sigma should recognize the following: (1) The structured and disciplined problem-solving approach of Six Sigma is an effective and efficient way of achieving a very difficult undertaking. (2) Because DMAIC is highly structured, there is higher consistency of project outcomes. (3) Because it is disciplined in making decisions based on data, the project results are sound.

The disciplined and structured approach of Six Sigma is especially effective in a transactional service environment. One of the more significant features differentiating transactional services from manufacturing is the absence of a tangible product. Applying a highly structured roadmap for studying intangible and unstructured processes increases the odds of success.

Another major difference between the two industries is that most manufacturing firms do not deal directly with the end-use customer — their customers tend to be distributors or wholesalers who add little value to the product other than that of easy access to the customer.[2] This again leads to the conclusion that a disciplined improvement approach that focuses on the customer is a good fit for an environment with constant direct contact with customers.

As we have seen over the past decade at Motorola, GE, and other companies, the Six Sigma approach has the potential to be effective in achieving the difficult goals of high-quality processes that meet the customers' needs profitably. However,

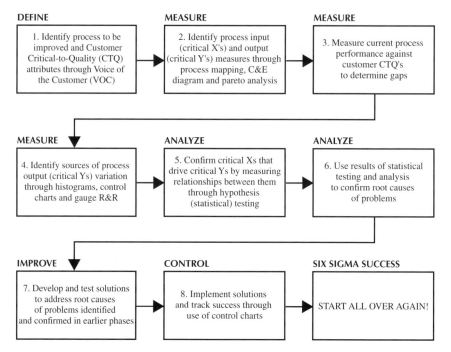

FIGURE 6.1 Eight steps to transactional Six Sigma success.

to reach the potential, appropriate application of the tools and techniques is necessary. Thus, the rest of this book covers how to effectively apply the appropriate tools to maximize the chance for success. Figure 6.1 provides a suggested path for reaching the goal of near-perfect processes; all of the tools along the path are described in this and the following chapters. Enjoy the ride.

6.2 DEFINE PHASE OVERVIEW

At this point in the Six Sigma process, projects have been selected and roles and responsibilities have been defined, as described in Chapter 2. The Define Phase is now ready to begin.[3] The four deliverables of the Define Phase are: (1) the project charter, (2) the VOC, (3) the preliminary process map, and (4) a fully formed and trained project team that is committed to accomplishing the project goals.

The project charter documents the Six Sigma project objectives, goals, estimated financial benefits, scope, and a preliminary timeline. The individuals assigned to the project, including the Champion, the Black Belt, the Green Belt (if applicable), and team members, are identified in the charter. VOC data are a key deliverable of the Define Phase. As stressed in Chapter 4, a sensible approach is to continuously gather VOC data to incorporate into the company's planning and operational processes. If this is done, the customers' CTQs are already known and used as a factor in project selection.

The preliminary process map is another key deliverable of this phase. A good tool for this is a S-I-P-O-C diagram, which describes the suppliers, inputs, process, outputs, and customers of the process. The preliminary process map provides a high-level representation of those involved in the process and the major activities that transform inputs into outputs.

The Define Phase is not complete until a fully formed, trained, and committed project team is in place. The team should be involved in finalizing the charter, including the project and communication plans.

As covered in Chapter 2, the type and amount of training for the project team varies according to the needs of the project; however, basic knowledge of the DMAIC process is essential. One half to one full day of training is usually sufficient for team members to gain understanding of the DMAIC process and the tasks required for successful project completion. The entire team, including the Black Belt and Green Belt, will continuously gain insight and knowledge as the project progresses.

In addition to the core Six Sigma team members, ad hoc team members, typically subject matter experts, need to participate at various times. This is especially critical in developing detailed process maps — those closest to the process must be involved in order to obtain true knowledge of the way the process really works on a day-to-day basis.

6.3 PROJECT CHARTER

The key elements of a Six Sigma project charter are the (1) business case, (2) problem statement, (3) goal statement, (4) roles and responsibilities, (5) scope of the project, (6) preliminary project plan, and (7) communication plan. General project information is provided in the heading of the charter, describing the project title, start and completion dates, and roles.

6.3.1 BUSINESS CASE

The business case is the rationale for doing the project — the compelling business reason to commit resources to the project. It states the importance of the project relative to the strategic goals of the company as well as the likely consequences if the project is not done. The business case needs to be defined by the project Champion, usually an executive-level individual who is directly responsible for the business line involved and thus has the most at stake in the success of the project.

6.3.2 PROBLEM STATEMENT

The problem statement describes why the project needs to be done. It answers the question, "What is the pain that caused this project to be selected?" A well-written problem statement is specific and includes the following elements:

- A concise statement of the problem
- Under what circumstances the problem occurs

- The extent of the problem
- The impact of the problem
- The opportunity presented if the problem is corrected

The problem statement *does not* describe causes or assign blame. A key point in the Six Sigma approach is fact-based decision making; root cause analysis at this point is probably based on opinion rather than fact.

6.3.3 GOAL STATEMENT

The goal statement describes the expected outcomes and results of the project. Like the problem statement, it is important to be specific in stating the goals. Improvement targets are described in quantitative terms, such as "a 50% reduction in average cycle time from 10 days to 5 days." The goal statement may change once the VOC requirements are better understood, because the goals should be directed at meeting those requirements.

Because Six Sigma is a business strategy rather than just a quality strategy, expected financial results are also included. Most companies set a minimum cost savings or revenue gain goal for a project to be selected for the DMAIC process. Estimates of the savings or gains and how they will be achieved through the project are provided.

The SMART criteria used to develop good measurements may be applied here to determine if the goal statement is well written. The goal should be:

- Specific
- Measurable
- Attainable
- Relevant
- Time bound

6.3.4 PROJECT SCOPE

The scope is one of the more important elements in the charter because it sets the boundaries of what is included and what is excluded. When defining project scope, seek a balance between being too broad, which diminishes the likelihood of success, and being too narrow, resulting in suboptimization of results.

The scope should be viewed as a contract between the Six Sigma team and the business units affected by the project. Otherwise, the business unit will tend to expand the scope midway through the project; this is known as scope creep.

6.3.5 PRELIMINARY PROJECT PLAN

The project plan for a Six Sigma study is often overlooked or included only as an afterthought. As covered in Chapter 5, project planning is critical to the success of a Six Sigma project. The project plan provides a roadmap of the major activities and tasks along with the responsible parties and expected completion dates.

Do not overcomplicate this with a sophisticated and complex planning tool, such as Microsoft Project, with tasks documented to the nth degree. Excel works just fine the majority of the time, although certainly Microsoft Project can be used if desired. The point is to use the project plan as a planning tool to ensure key milestones are met; avoid letting it become overly burdensome.

6.3.6 COMMUNICATION PLAN

Successful orchestration of a Six Sigma project is difficult at best. From the Champion to individual team members, and everyone in between, coordination of roles, responsibilities, and expectations is crucial, given the relatively short time frame of most Six Sigma projects (4 to 6 months). A communication plan is the tool to ensure this coordination. A communication plan addresses the following questions:

- Who is responsible for updating and delivering the communication plan?
- What critical content will be distributed (e.g., project charter, project plan, meeting minutes)?
- How will the information be distributed? For each critical content item listed on the plan, list the mechanism of communicating it, such as formal presentations or memos, e-mails, etc., and the person responsible for delivery of the content item.
- When is each content item to be distributed (e.g., the frequency of project plan updates)?
- Who is to receive the information (i.e., the distribution list for each content item)?

The communication plan will likely need updating once the project is underway — there are always individuals left off the distribution list that you will hear from at some point in the project.

6.3.7 UPDATING THE CHARTER

Much of the information documented in the project charter is preliminary, based on the best available data. If all the data required to understand the project results are known at this point, chances are it is not really suited for a Six Sigma DMAIC approach. Such a situation is often referred to as an IC (Implement and Control) project (pronounced "ICK"), because the solution to the problem is known and it is a matter of implementing and controlling the results.[4]

The elements most likely to change based on updated information are the goal, preliminary project plan, and communication plan. The business case and the problem statement are likely to remain as is because they are the key drivers of the project. Of course, business strategies do change, so it may be necessary to revisit these elements. However, anything but minor revisions here requires major changes to the other elements and thus to the very nature of the project.

6.4 VOICE OF THE CUSTOMER

Chapter 4 described the importance of accurately capturing the VOC. Business success depends on providing customers what they want, unless of course your business is a monopoly. The concept is deceptively simple and the execution of it deceptively difficult.

The best-case scenario is that VOC is known and understood at this point in the DMAIC process due to continuous measurement and feedback of customer requirements. VOC then drives project selection, and capturing customer requirements during project definition is minimal. However, this is the exception rather than the rule. Even if the customer requirements are known beforehand, specific CTQ attributes may need further definition in the Define Phase.

Gathering customer CTQ requirements is necessary to ensure processes are designed to meet those requirements. However, customer feedback, whether through formal surveys or complaint data, is usually not in the form of specific requirements needed for process design (e.g., "I want my breakfast delivered to my room no more than 30 minutes after placing the order."). It is more likely to be vague, such as "It takes too long to get breakfast delivered." The job of the Six Sigma team is to translate the vague feedback into measurable requirements, often through further questioning of the customer, using specific language and examples (see Figure 4.2).

CTQs are different for a transactional environment than for manufacturing. Transactional CTQs are more likely to be intangible, such as responsiveness, courtesy, competence, accuracy, and timeliness. In terms of perceived quality of a process, timeliness is a key driver — think about the dissatisfaction you experience when you have to wait a long time on hold or in a long queue to receive service. Even if every other aspect of the transaction is flawless, the frustration from excessive waiting dominates the entire experience. For example, to develop a specific timeliness measure from VOC data, the statement "It takes too long to get breakfast delivered," is translated into a specific and measurable CTQ, such as "breakfast delivered within 30 minutes after ordering."

One of the first steps in identifying customer CTQs is to identify the customers of the process under consideration. Although the focus should be on the paying customer, it is important to consider other "customers" of the process. A customer is defined as anyone receiving output from the process. It is easy to see that many processes have internal customers as well as external customers.

Once key customers have been identified, conduct research to capture each customer segment CTQ requirements. From passive analysis of complaint data to active surveying, the deliverable of the research is a list of customer needs, which is then translated into specific and measurable CTQ requirements. Refer to Chapter 4 for more information on capturing VOC.

6.5 HIGH-LEVEL PROCESS MAP

The third key deliverable of the Define Phase is a high-level process map to ensure clarity is reached on the project definition. The project team should develop the map, with the understanding that it is a starting point in gaining an in-depth knowledge of the current process and its performance.

A useful tool in developing the process map is a S-I-P-O-C diagram. Detailing the suppliers, inputs, process steps, outputs, and customers promotes a common understanding of the project scope (where does the process begin and end?), the inputs and outputs of the process, and the customers who receive the outputs. If VOC data is available, the requirements of the customer should be added to the diagram, resulting in a S-I-P-O-C-R.

The completed S-I-P-O-C diagram needs to be shared with the project Champion and other key stakeholders to ensure the project team is heading in the right direction. It is far better to clear up any misunderstandings at this point in the project than wait until substantial work is completed; the project charter and S-I-P-O-C are the appropriate tools to accomplish this goal.

The S-I-P-O-C diagram documents the current process. It may seem easier to instead start describing the process as it *should* be; however, this is incorrect and counter to the Six Sigma approach. If improvements to the process are known at this point, why go through the Measure and Analyze Phases of the DMAIC process? Factual analysis, rather than opinion, is the basis for improvements, and improvement solutions at this point are usually opinion. The steps involved in developing a S-I-P-O-C diagram are:

1. Outline the major activities involved in the process. There should be four to six high-level steps, drawn in block diagrams. An action verb is used as the first word to describe the process steps, such as "Download file from system" or "Review application." Recall from Chapter 3 that a process transforms an input into an output. Describing the major steps involved in this transformation is the goal here.
2. Identify the outputs of the process. There may be more than one output, from one or more of the process steps. Outputs are both tangible, such as a billing statement, and intangible, such as information relayed through a phone call.
3. Determine the inputs into the process (e.g., a completed application or information received through e-mail).
4. Identify the suppliers of the inputs (e.g., the source of the application or the e-mail). It is common that the suppliers are also the customers of the outputs.
5. Document the requirements of the customer, if they are available. A note of caution is warranted here — if the customer's requirements have not been researched and translated into CTQs, they should not be included at this point. It is incorrect to document customer requirements based on opinion. It is tempting, especially for long-term employees, to believe they have an understanding of the customer's requirements. Although they may be correct, it is critical that their assumptions be validated. A distinguishing characteristic of Six Sigma is a reliance on the scientific method of data-driven and systematic research to draw conclusions.

The purpose of creating the S-I-P-O-C diagram is to develop a high-level representation of the process being studied; it is not meant for detailed process analysis.

One of the first tasks in the Measure Phase is to develop a detailed process map, which is the appropriate tool for gaining in-depth knowledge of the process.

6.6   PROJECT TEAM

Significant improvements to key processes can rarely be achieved through the work of only one or two individuals. A contributing factor to the success of the Six Sigma approach is placing responsibility for developing and implementing process improvements on the employees actually involved in the process, rather than on quality experts in the quality department. Although Black Belts are usually full-time positions dedicated to improvement projects, their role is to lead the Six Sigma project team toward solutions rather than develop solutions on their own. Successful implementations of changes depend on employees buying into and getting involved in the process, perhaps more than the quality of the solution itself. A team-based approach involving key stakeholders of the process ensures buy-in. Moreover, the quality of the solution is likely to be higher when those closest to the process are involved in developing the improvements. The Japanese caught on to this fact almost half a century before it was standard practice in Western quality improvement efforts.

As covered in Chapters 5, getting the right composition of team members is critical to the success of the project. There are guidelines to follow when selecting team members; however, they are merely guidelines. The nature of the project determines to a large degree the team member composition. For example, a process heavily dependent on a particular automated application suggests experts in the application should be on the project. A project to reduce the cycle time of internal mail delivery may not need systems representatives. Common sense dictates that representatives from departments who have a major role or stake in the project outcome should be included as core team members. Subject matter experts can be brought in as needed from areas that have a peripheral role. A question that often arises is whether the customer should be on the team, especially internal customers. Again, it depends on the nature of the project, but in general, the answer is yes, because the idea is to design processes to meet customer requirements. As an active participant, the customer as a team member is an excellent source to ensure requirements are understood and met.

The one rule that should be consistently followed is that employees involved in carrying out the work on a day-to-day basis should be involved. Those closest to the process are in the best position to accurately document the current workflow and understand if a particular solution is feasible under real-world conditions. Often, the difference between management's understanding of a process and the way the work is actually accomplished is truly surprising.

The size of the core Six Sigma team should be limited to no more than seven. Anything over this may lead to confusion and time delays in completing the project. However, other employees should participate in a supportive role at various times throughout the project. As covered in Chapter 5, the criteria for selecting team members includes enthusiasm toward changes, analytical ability, dedication, unique knowledge or skill of a key aspect of the process, and a personal style oriented toward teamwork. There is no room for power-hungry individuals who view the Six Sigma project as a way to further their own agenda. Cooperation and a positive view of constructive change that benefits the bottom line are required for project success.

6.7 CASE STUDY: DEFINE PHASE

The case study used to illustrate Six Sigma tools and techniques is reducing call volume into a 400-seat inbound call center that handles disputes and inquiries for a major credit card company known as Card One. Card One, based in New York, has a total card portfolio of 9.2 million U.S. customers. Two operations centers handle inbound calls and make outbound collection calls: one in Atlanta and the other in Omaha. A third site in Dallas processes card applications and handles merchant operations.

The Six Sigma project focuses on reducing the number of inbound calls regarding inquiries and disputes. Although the project team is based in Atlanta, results will be shared with the 350-seat Omaha site because their operations are very similar. The Atlanta facility receives an average of 800,000 calls per month and the Omaha facility 730,000 calls per month, for a monthly average of 1.53 million and an annual average of 18.36 million inbound calls. The goal of the project is to reduce the number of inbound calls by 10%, or 1.836 million calls per year. At an average cost of $3.62 per call, the savings are estimated at almost $6.65 million.

An ongoing process exists for capturing the VOC through quarterly transaction-based surveys. The surveys are sent to a random sample of 1,000 cardholders who called in to the center in the previous 3 months. There is a 28% response rate, resulting in approximately 280 observations per quarter, or 1,120 per year. The transactional surveys measure the customer's satisfaction with the particular transaction that the customer had with the company, in addition to ongoing transactions such as billing and payment processes. (See Figure 4.1 for a sample survey.)

The project charter, preliminary project plan, and communication plan for the case study are shown in Figures 6.2 through 6.4. The S-I-P-O-C diagram for the case study is presented in Figure 6.5. Preliminary VOC results were provided in Figure 4.3.

Many of the examples in the next four chapters (Measure, Analyze, Improve, and Control Phases) relate to the case study presented here. Chapter 11 on DFSS and Chapter 12 on Lean Servicing continue with the theme of improving operations at Card One, but for different processes.

Card One Six Sigma Project Charter	
Project Information:	Resource Information:
Project #: 1201	Master Black Belt: Jackie Honeycutt
Project Name: Inbound Call Volume Reduction	Black Belt: Michael Antonio
Project Start Date: February 4, 2002	Green Belt: N/A
Projected End Date: August 30, 2002	Champion: Paul Morris, V.P. Customer Service
Team Members: Larry David, Director, Customer Service, Leah Rich, Director, Human Resources, Tara McNamara, Manager, Customer Service, Daniel McCray, Manager, Collections, Carl Friedman, Supervisor, Customer Service, Andrew Jacobs, Correspondence Supervisor and Adam Benjamin, Customer Service Representative	

Business Case: The number of inbound calls into the Customer Service 800# has increased proportional to the number of cards in force in the past three years. The average calls per card per year have increased from 1.6 in 1999 to 2.0 for 2001, resulting in additional costs of $5.4 million. If the increase in calls to card volume continues at a similar rate, the company's profitability and share price will be negatively affected.

Problem Statement: The number of inbound calls into the operations center continues to increase relative to the number of cards in force (CIF). There is no apparent reason for the increase since the card features and benefits have remained relatively stable.

Goal Statement: The goal of the project is to reduce the number of inbound calls by 10%. Based on 2001 figures, this would equate to a reduction of 1,836,000 calls per year, resulting in annual savings of $6.65 million.

Project Scope: The scope of the project is to examine all processes relating to call reasons for cardholders calling into the 800# for inquiries and disputes. The focus will be on the relatively few processes that result in the most calls. The scope includes examining the process for handling inbound calls and the follow-up processes required to close out inquiries and disputes, as well as the management organization in the call center.

The scope does not include staffing and scheduling system/processes. The inbound calls into the Dallas merchant operations center are not included in the scope.

Preliminary Project Plan: See Attachment A

Communication Plan: See Attachment B

FIGURE 6.2 Case study project charter.

	Attachment A: Card One Six Sigma Project Charter Call Volume Reduction Project Plan			
	Project #: 1201			
	Projected Completion Date: August 30, 20XX			
	Version: 1.0			Page 1
Task #	Task Description	Responsible	Due Date	Comments
1.0	Define Phase	Team	2/22	
1.1	Develop first draft of project charter, including project plan and communication plan	P. Morris M. Antonio	2/4	
1.2	Hold project kick-off meeting	M. Antonio	2/8	Meeting scheduled for Wed, 2/6
1.3	Finalize Project Charter, including project plan and communication plan	Team	2/13	
1.4	Develop method to capture and document Voice of the Customer	L. David C. Friedman	2/13	Quarterly transaction surveys are primary source of VOC
1.5	Develop preliminary process map/SIPOC	M. Antonio Team	2/20	Meeting scheduled for Mon 2/18
1.6	Formally document VOC research results	L. David C. Friedman	2/20	Any follow-up work needs to be identified if VOC is not completed by due date
1.7	Review results of Define Phase with Sponsor and MBB	Team P. Morris J. Honeycutt	2/22	Michael to schedule meeting

FIGURE 6.3 Case study preliminary project plan.

Attachment B: Card One Six Sigma Project Charter Call Volume Reduction Communication Plan
Project #: 1201
Projected Completion Date: August 30, 20XX
Version: 1.0   Page 1 of 1

What	Responsible	Distribution	Frequency	How	Comments
Project Plan	M. Antonio	Team P. Morris J. Honeycutt	Bi-weekly	Via e-mail	Plan will be updated at weekly team meeting
Team Meetings	M. Antonio	Team	Weekly	Invitation via e-mail calendar	Meetings to be held in first floor conference room
Meeting Minutes	Assigned Scribe	Team	Weekly	Via e-mail	Scribe will be rotated among team members
Electronic Working Documents	Team	Team	As needed	Shared Six Sigma drive in project folders	All project documentation is to be stored on shared drive – please date and indicate version number of document
Hard Copy Documents	M. Antonio Team	Master Project Binder stored in Michaels office	As needed	Master Project Binder	Final documentation kept in Master Project Binder. Team responsible for providing to Michael for inclusion
Review Meetings	M. Antonio	Team P. Morris J. Honeycutt	At end of each phase and as needed	Invitation via e-mail calendar	Review meetings separate from weekly team meetings. All meetings held in first floor conference room.
General Company Update	T. McNamara	All company employees	Monthly	Company Newsletter	Tara responsible for writing project status summary for inclusion in company newsletter

FIGURE 6.4 Case study communication plan.

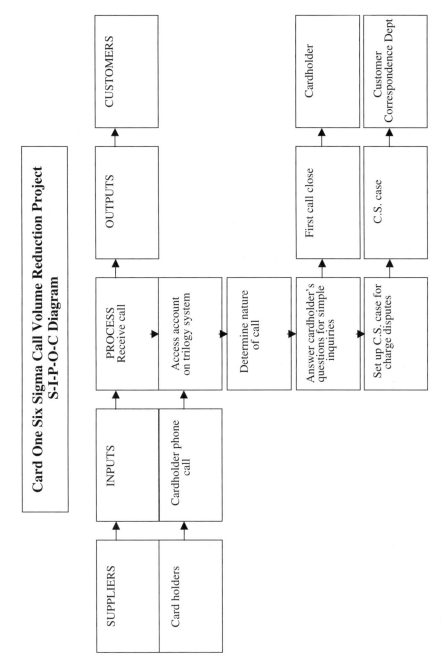

FIGURE 6.5 Case study S-I-P-O-C diagram.

REFERENCES

1. Noguera, J. and Nielson, T., "Implementing Six Sigma at Interconnect Technology," ASQ FAQ Series on Six Sigma, 2000, from a presentation at the 46th Annual Quality Congress, 1992.
2. This is rapidly changing through the growth of e-commerce, which facilitates the sales exchange from producer directly to consumer.
3. Most of the early implementations of the Six Sigma methodology did not include the Define Phase; MAIC was the accepted approach during that time. It appears that GE was the first to include the D in DMAIC in the mid-1990s, and others slowly caught on. Some Six Sigma consulting and training firms still exclude the D in DMAIC, preferring to define the project in the project selection process.
4. Pande, P.S., Neuman, R.P., and Cavanagh, R.R., *The Six Sigma Way: How GE, Motorola, and Other Top Companies Are Honing Their Performance*, McGraw-Hill, New York, 2000.

7 Measure Phase

7.1 OVERVIEW

Objectives of the Measure Phase include identifying process performance measures and setting their targets according to the VOC data. The current process is then evaluated against the targets. The gap between current and target performance provides the Six Sigma team with priorities and direction for further analysis.

Identification of the process inputs and how they affect the process outputs are initiated during this phase. This involves identifying appropriate measurement data, followed by collecting the data and using it to measure baseline performance. Activities of the Measure Phase are:

1. Determine appropriate process performance measurements, such as timeliness and accuracy measures, from an understanding of the customer and business CTQs.
2. Measure baseline sigma levels, which is current performance against the target process performance measure, in order to determine the capability of the current process to meet customer CTQs. This is referred to as gap analysis.
3. Define and measure process defects.
4. Develop a detailed process map of the current process.
5. Conduct best practices and benchmark research results, including how the current process is performing against best practices.
6. Identify the process inputs (critical X's) and outputs (critical Y's) and their relationship to each other.

The work involved in the Measure Phase is substantial and can take many directions, depending on the nature of the project. If the primary project goal is to reduce the cycle time of a process, the Measure Phase will probably focus more on process mapping and collecting data related to the cycle time. If the goal is general improvement of a process, relating to various dimensions of accuracy, costs, cycle time, and customer satisfaction, then the Measure Phase may be more data intensive. Proper scoping of the project during the Define Phase is necessary to provide the project team with direction on where to focus their data collection and measurement efforts during the Measure Phase. The most common Six Sigma tools used during the Measure Phase are process maps, check sheets, Pareto charts, cause-and-effect (C&E) diagrams, histograms, and control charts. Statistical techniques to understand process data are also used in the Measure Phase.

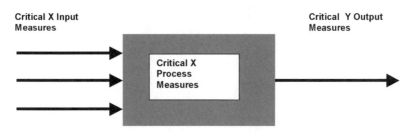

FIGURE 7.1 Input, process, and output measures expressed as critical X's and Y's.

7.1.1 Types of Measures

One of the first steps in the Measure Phase is to define the performance measures *from the customer's perspective*. For cycle time reduction projects, it is logical to start with measuring the current cycle time of the process against customer timeliness requirements. For example, if you are measuring the cycle time to respond to written inquiries, the time from when the customer mails the letter to the time the customer receives the response is the appropriate cycle time measurement. Although the time spent going through the postal service system may not be entirely under your control, it is irrelevant from the customer's perspective — customers usually do not distinguish between the days spent in the postal system and the days spent within the company's internal environment.

For general improvement projects, defining appropriate measures may be a little more difficult. The three types of measurements needed are output, process, and input measures, described further below. The Input-Process-Output components of the S-I-P-O-C diagram from the Define Phase are useful in identifying these measures, as indicated in Figure 7.1.

Input measures: These are usually volume driven, such as the number of cards in force, number of inbound calls, and type of call. They are referred to as critical X's or independent variables.

Process measures: These are also called critical X's or independent variables. Business CTQs are typically process measures. Recall from Chapter 3 that a process is a series of definable, repeatable, and predictable interrelated tasks through which inputs are transformed into outputs. Process measures relate to the effectiveness and efficiency of the resources necessary to transform inputs into outputs, such as labor hours, training, and equipment. *Effectiveness* is defined as the ability of the process to provide the desired outcome. *Efficiency* is providing the desired outcome utilizing as few resources as necessary. In the call center study, process measures include average handling time, percent agent adherence to schedule, transfers, and unit cost.

Output measures: These are customer CTQ attributes, identified through VOC. They are referred to as critical Y's, dependent variables, or response variables. Output measures for a service process typically relate to timeliness,

Continuous	Discrete
Length	Proportions
Weight	Defects
Time Duration	Yes/No or Pass/Fail
Temperature	A, B, C, D Grades or a 5-point survey scale

FIGURE 7.2 Continuous vs. discrete data.

accuracy, access, and responsiveness to customers' needs. In the Card One case study, a major CTQ and output variable is first-call resolution, which relates to the timeliness, accuracy, and responsiveness of the transaction. Research indicates that customers do not want to have to call more than once to have a question answered or a problem resolved.[1] Each subsequent call back results in significant decreases in satisfaction. Other types of output measures in the call center include the time it takes for a customer to reach an agent (access and timeliness measure), knowledge of the agent (accuracy measure), and whether the transaction left the caller feeling like a valued customer (responsiveness measures). How well the company does on each of these measures determines overall satisfaction — an important output measure in any situation.

At this stage in the project the process and input measures are probably identified only as *potential* critical X's. Only through the Six Sigma process of systematic measurement and discovery are the actual critical X's identified. Through application of the tools in the Measure and Analyze Phases, the potential critical X inputs prove either critical or not to the critical Y outcomes.

7.1.2 MEASUREMENT DATA TYPES

Measurement data are categorized as either continuous or discrete (Figure 7.2). *Continuous* data, also called variable data, are based on a continuum of numbers, such as heights, weights, times, and dollars. *Discrete* data, also called attribute data, are based on counts or classes. If only two options exist — flipping a coin to yield either a head or a tail or flipping a switch to either on or off — they are referred to as *binomial* discrete data. Discrete data may also involve more than two classes, such as the A–F grading scale or filing customer complaints into six categories.

Continuous data are preferred over discrete data primarily because they provide more information. Continuous measurements can be graphed and charted over time as well as provide greater understanding of the underlying process. For example, to determine whether you are meeting the CTQ of easy access to a phone agent, the preferred measure is how many seconds it took for the agent to answer (continuous), not whether the agent answered in less than 30 sec (discrete). The advantages of measuring the exact number of seconds rather than a yes/no criterion are a better understanding of the actual process and, more importantly, how capable the process is in meeting customer requirements.

7.2 INTRODUCTION TO STATISTICAL METHODS

7.2.1 DESCRIPTIVE VS. INFERENTIAL STATISTICS

Descriptive statistics *describe* characteristics about a population or sample of data, whereas *inferential* statistics is concerned with *inferring* characteristics of the population from a sample of data. Descriptive statistics is interested in "what is"; inferential statistics is interested in "what will be," or extrapolating from the sample to the population. The statistics utilized in Six Sigma projects are primarily inferential, because we are usually interested in understanding the total population data based on sample data.

Inferential statistics are used mainly for two purposes: estimation and hypothesis testing. *Estimation* involves using sample data from a population to estimate a value from that population, such as the average. Because we are taking a sample of data rather than the entire population, we must make judgments about how confident we are that the sample values estimate the population values.

The second use of inferential statistics, *hypothesis testing*, involves testing a claim about a set of data against an alternative claim. One form this may take is testing the strength of a relationship between two variables (correlation analysis). A second form this may take is hypothesizing that a sample of data belongs to a particular population vs. an alternative hypothesis that the sample of data does not belong to that population. An example is what is known as a two-sample *t* test, in which we need to determine if two different samples belong to the same underlying population. The first hypothesis would say, yes, these two samples belong to the same underlying population (with similar means and variances), and the alternative hypothesis would say, no, these two samples are different enough that we cannot be confident that they came from the same population and thus a significant difference exists. The way we determine how confident we are in our estimations and hypotheses is through tests of confidence. Applications of these tests and others are explained in Chapter 8.

7.2.2 BASICS OF PROBABILITY AND STATISTICS

Understanding statistics starts with an understanding of probability concepts. To understand probability, let us start with an example of categorizing continuous data into discrete class intervals. The example used is tabulating data on the years of registered nursing experience of RNs, calculated from a random sample of all RNs in the U.S. (Table 7.1). The sample of 440 responses is defined as all RNs employed in the profession for at least 1 year. The data are categorized in five distinct class intervals, representing years of experience, and the frequency of each class. This is called a *frequency distribution*. If we convert the frequencies into percentages, we have a *percentage distribution*, also called a *relative frequency distribution*, because the percent frequency in each class is relative to the total number of frequencies for all classes (Table 7.2). The graph of the results provides a histogram of the data distribution and shows the relative class frequencies of years of experience for RNs (Figure 7.3).

TABLE 7.1
Frequency Distribution of RN Years of Experience

Years of RN Experience	Frequency (Number of Nurses in Each Experience Class)
1–5 years	89
6–10 years	120
11–15 years	108
16–20 years	78
21–40 years	45
	440

TABLE 7.2
Relative Frequency Distribution of RN Years of Experience

Years of RN Experience	Relative Frequency (Percentage of Nurses in Each Experience Class)
1–5 years	20%
6–10 years	27%
11–15 years	25%
16–20 years	18%
21–40 years	10%
	100%

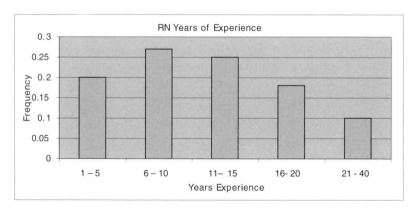

FIGURE 7.3 Histogram of relative frequency distribution of RN years of experience.

Probability is defined as the long-run relative frequency with which an event, or outcome, will occur. In the RN example, an outcome is the class interval associated with the number of years of experience. Five possible outcomes correspond to the

---

TABLE 7.3
Relative Frequency Distribution of RN
Years of Experience

Years of RN Experience	Probability
1–5 years	0.20
6–10 years	0.27
11–15 years	0.25
16–20 years	0.18
21–40 years	0.10
	1.00

---

five class intervals of years of experience. Each outcome is defined as a *random variable*. Define the random variable x as follows:

$x = 0$ for 1 to 5 years experience
$x = 1$ for 6 to 10 years experience
$x = 2$ for 11 to 15 years experience
$x = 3$ for 16 to 20 years experience
$x = 4$ for 21 to 40 years experience

The word "random" means that the variable will assume its values in a chance (probabilistic) manner. As long as the values of the variables are equally likely to occur, then it is reasonable to assume randomness. The word "variable" means that we do not know in advance what value the outcome will take.

Converting the relative frequency distribution to probabilities of occurrence provides a *probability distribution*, which is illustrated in Table 7.3. The graph in Figure 7.4 provides a probability density function (PDF) from the relative frequencies. What does this mean? A PDF is a mathematic model of the frequency distribution — it provides the probability that a particular value (outcome), x, falls within the area under the curve between two distinct points. In the RN example, what is the probability that a randomly chosen RN will have between 6 and 10 years of experience, or $x = 1$? From the graphs, we can answer that this probability is 0.27.

The PDF for discrete values is different from the function for continuous data. Discrete data can be counted and expressed as whole integers. Continuous data is based on a continuum of numbers that can be broken down into smaller and smaller values, tending toward infinity. In the RN example, continuous data (units of time) were converted into discrete outcomes by categorizing the data into discrete (countable) class intervals, with which we can associate probabilities. The following provides a more detailed explanation:

Discrete probability distributions: A discrete probability function $p(x)$ of the discrete random variable x is the probability $p(x_i)$ that x can take the specific value x_i. In the RN example, this is equivalent to saying that the probability of the random variable x_1 (or $x = 1$, translating into the outcome of 6–10

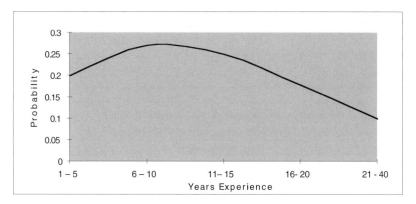

FIGURE 7.4 Probability density function of RN years of experience.

years experience) is 0.27. The PDF is nothing more than the distribution, or portrayal, of the probabilities of occurrence of each value (or interval of values) of the random variable. The probability of occurrence of any one random variable x, or possible outcome, is between 0 and 1, because there is no concept of a probability of an outcome being less than zero or greater than 100%. Finally, the sums of all the probabilities of x must equal 1.

Continuous probability distributions: Because continuous data can be broken down into smaller and smaller units tending toward infinity, the probability that a random variable, x, can take on a single value, x_i, is 0. What we measure is the probability that a random variable, x, will assume a value between two intervals. The continuous distribution that Six Sigma is most interested in is the *normal distribution*, which is covered in Section 7.2.4.

7.2.3 PROBABILITY DISTRIBUTION MEASUREMENTS

The two most important characteristics to describe most data distributions are measures of central tendency and dispersion. Central tendency measures include the mean value, the median, and the mode. Dispersion measures include the range, standard deviation, and variance.

7.2.3.1 Central Tendency Measures

Central tendency measures, also called measures of location, characterize the center or middle area of the data. The *mean* value is the average value of all data points and is the most common measurement of data in business applications, including Six Sigma. It is derived through summing all the data points and dividing by the number of data points. The second measure of central tendency is the *median*, which is the midpoint of the data when it is sorted in ascending order. The median is the value at which one half of the data points are below and the other half above. If the number of data points is odd, the median is the central data point. If the number of data points is even, the median is calculated by averaging the two middle points. The third measure of central tendency is the *mode*, which is the value that occurs

Observations: [22, 25, 28, 29, 31, 32, 32, 34, 38, 41, 44, 47, 51, 56]

Mean: add all observations and divide by number of observations = [(22 + 25 + 28 + 29 + 31 + 32 + 32 + 34 + 38 + 41 + 44 + 47 + 51 + 56)/14 = 36.4]

Median: List all values from low to high and take the middle value if number odd number of observations or average the two middle values if even = [(32 + 34)/2 = 33.0]

Mode: 32

Range: Subtract lowest observation from highest = [56 – 22 = 34.0]

Variance:

1) Subtract each observation from the mean value = [–14.4, –11.4, –8.4, –7.42, –5.4, –4.4, –4.4, –2.4, 1.6, 4.6, 7.6, 10.6, 14.6, 19.6]

2) Square each of the differences: [208.2, 130.6, 71.0, 55.2, 29.5, 19.6, 19.6, 5.9, 2.5, 20.9, 57.3, 111.8, 212.3, 383.0]

3) Sum the squares and divide by the number of observations = [1327.4/14 = 94.8]

Standard Deviation: Calculate the square root of the variance = $\sqrt{94.8}$ = 9.74

FIGURE 7.5 Calculation of location and dispersion measures of class age data.

most often. Each of these measures is calculated in Figure 7.5, which provides population data on the age of a group of 14 attendees in a Six Sigma Black Belt class.

7.2.3.2 Dispersion Measures

Although the central tendency measures provide important information about the center of the data, they do not provide enough information to determine how the data are distributed around the central point or their variability. This is an important point in Six Sigma. In most business environments, the focus is on the average value, without a clear understanding of the distribution of the data. Although the average value is necessary to understand aspects of the data, it does not take into account extreme values. We could say that, on average, the time to reach a call center agent is 30 sec, but a look at the distribution of the data tells us that the range is from 2 to 240 sec — two very different scenarios. Measures of dispersion help us identify how data values are distributed and, together with the mean, provide the information we need to calculate probabilities and sample sizes for most distributions.

The *range* is calculated by subtracting the lowest value in the data set from the highest value. This measures the distance between the two extreme values but provides no information on the central tendency of the data. The *deviation* of any one data point is a measure of the distance of that point from the mean value. The *average deviation* is calculated by summing the deviations of all the data points from the mean and dividing by the number of observations (*n*). However, this is not a very useful measure because it will always equal 0 — the positive deviation scores are always canceled out by the negative scores. To solve this problem, we square the deviations, leaving all positive numbers, then take the average of the squared deviations by summing them and dividing by *n*. This provides a measure of the *variance* of the data, as illustrated in Figure 7.5.

One issue with using the variance to describe dispersion is the variance is a unit of measurement that has been squared, rather than the original non-squared units. In Figure 7.5, the variance is a measure of age squared — not a very useful unit of measure. Because of this, we need to take the square root of the variance in order to get back to the original unit of age. When we take the square root of the variance, we arrive at the *standard deviation*, which is the most widely used dispersion statistic in Six Sigma.

7.2.4 THE NORMAL DISTRIBUTION

One of the most useful distributions in Six Sigma, and statistics in general, is the *normal distribution*. The normal distribution is bell shaped and symmetrical, with most values around the mean and similar values on either side of the mean. Normal distributions are a family of distributions that have the same general shape. Figure 7.6 provides examples of normal distributions. Note the difference in the spread of the curve for each diagram. However, the area under each curve is the same.

Describing a normal distribution requires only two numbers: the mean, μ, and the standard deviation, σ. For sample data, the two numbers are the sample mean, X, and the sample standard deviation, S. The normal distribution is representative of a large number of naturally occurring values, such as height, weight, age, test scores, and many others.[2]

The *standard* normal distribution is a normal distribution with a mean of 0 and a standard deviation of 1. Normal distributions can be transformed to standard normal distributions by the following formula:

$$Z = \frac{X - \mu}{\sigma} \tag{7.1}$$

where X = the variable of interest from the original normal distribution, μ = the mean of the original normal distribution, and σ = the standard deviation of the original normal distribution. The Z score reflects the number of standard deviations above or below the mean of a particular X. If, for example, a person scored a 75 on a test that has a mean of 70 and a standard deviation of 5, then they scored 1 standard deviation above the mean. Converting the test scores to Z scores, an X of 75 would be:

$$Z = \frac{75 - 70}{5} = 1.0 \tag{7.2}$$

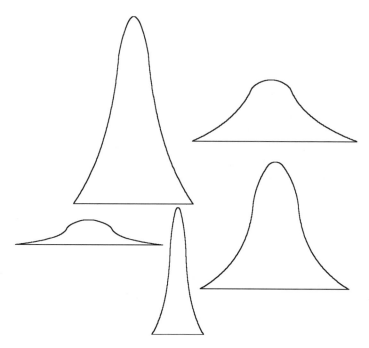

FIGURE 7.6 Examples of normal distribution with different locations and dispersion statistics.

A Z score of 1 means the original score was 1 standard deviation above the mean; $Z = 2$ means it was 2 standard deviations from the mean; and so on. Applying the formula $[Z = (X - \mu)/\sigma]$ will always produce a transformed variable with a mean of 0 and a standard deviation of 1. However, the shape of the distribution will not be affected by the transformation. If X is not normal then the transformed distribution will not be normal either. Percentages under portions of the standard normal distribution are shown in Figure 7.7. About 0.68 (±0.341) of the distribution is between −1 and +1 standard deviations from the mean; about 0.96 (±0.477) of the distribution is between −2 and +2 standard deviations; and 0.997 (±0.498) is between −3 and +3.

One important use of the standard normal distribution is converting scores from a normal distribution to a percentile rank. For example, if a test is normally distributed with a mean of 70 and a standard deviation of 5, what proportion of the scores are above 75? This problem is similar to figuring out the percentile rank of a person scoring 75. The first step is to determine the proportion of scores that are less than or equal to 75. This is done by figuring out how many standard deviations above the mean 75 is. Because 75 is 5 points above the mean (75 − 70 = 5) and the standard deviation is 5, a score of 75 translates to 5/5 = 1 standard deviation above the mean. Or, in terms of the formula,

$$Z = \frac{X - \mu}{\sigma} = \frac{75 - 70}{5} = 1.0 \tag{7.3}$$

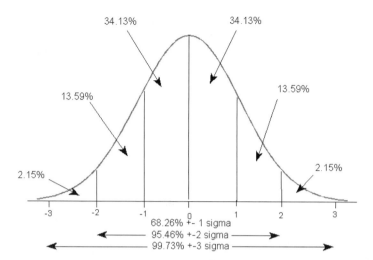

FIGURE 7.7 Percentage of data observations within 1, 2, and 3 standard deviations from the mean for normally distributed data.

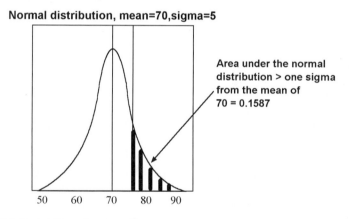

FIGURE 7.8 Probability of test scores greater than 75.

The Z table in Appendix B.1 can be used to calculate that 0.841 of the scores are less than or equal to a score 1.0 standard deviation above the mean. It follows that only 0.158 of the scores ($1 - 0.841 = 0.158$) are above a score 1.0 standard deviation above the mean. Therefore, only 0.158, or 15.9%, of the scores are above 75 (Figure 7.8).

This section has provided an overview of the very important normal distribution. Section 7.8 describes other continuous distributions along with common discrete distributions used in Six Sigma projects.

Formula	Population Parameter Formulas	Sample Statistic Formulas
Mean	$\mu = \dfrac{\sum (x)}{N}$	$X = \dfrac{\sum (x)}{n}$
Standard Deviation	$\sigma = \dfrac{\sqrt{\sum (x_i - \mu)^2}}{N}$	$S = \dfrac{\sqrt{\sum (x_i - \text{mean}[x])^2}}{(n-1)}$

FIGURE 7.9 Population parameter formulas vs. sample statistic formulas.

7.2.5 POPULATION PARAMETERS VS. SAMPLE STATISTICS

Recall from earlier in this section the population of 14 Black Belt training attendees. We were interested in describing only the measures of central tendency and dispersion for this group only, with no plans to extrapolate to other class groups. Recall also the discussion on the difference between descriptive statistics and inferential statistics. Descriptive statistics describe what is, based on either the population or a sample. Inferential statistics allow us to infer characteristics of the population through measuring characteristics of the sample. Because we have designated that the ages of the 14 attendees is the population, we can make no statements about the ages of other Six Sigma class attendees based on our measurements. The measures of central tendency and dispersion we calculated are referred to as *population parameters*. The population mean of 36.4 is denoted by μ and the standard deviation is denoted by σ. If we wanted to infer characteristics of the population of Six Sigma class attendees, we would use slightly different formulas to measure the sample mean and standard deviation — we refer to these measurements as statistics, or *sample statistics*. The notation for the sample mean is X and the sample standard deviation is S. Using the sample of 14 observations to make inferences about a larger population requires that the sample is *randomly* chosen from the larger population; otherwise, the sample statistics do not apply. Figure 7.9 shows the formulas for the population and the sample mean and standard deviation. The sample statistics are used to extrapolate to the larger population of all Six Sigma class attendees. We can also state the uncertainty in our sample estimates (e.g., how far the sample mean is from the population mean) through the use of confidence intervals and confidence levels, described in the next section.

7.2.6 SAMPLING PLAN

Advances in database and reporting tools in recent years have made it more practical to collect population data rather than a sampling of the data. This is especially true in call centers, where data on every call are captured and sophisticated reports are produced that offer an abundance and often overwhelming amount of information on call transactions. However, sampling is usually more economical, resulting in the need to develop a sampling plan.

Sampling involves collecting data on a subset of the population data in order to make inferences about the population measures. As discussed earlier, statistical inference is the science that provides the tools necessary to estimate the confidence we have in using a subset of the data to draw conclusions about the entire population. When we estimate population parameters, such as mean and standard deviation, using a sample of data, we call that a point estimate, or *statistic*.[3] Clear documentation of what the data will be used for is necessary before actual design of the sampling plan. Assuming this has been done, the steps to develop a statistics-based sampling plan are:

1. Define the target population.
2. Select a sampling scheme, which is a detailed description of what data will be collected and how.
3. Determine sample size.

These steps are described below.

7.2.6.1 Defining the Target Population

Although defining the target population may seem trivial, choosing the wrong population from which to draw samples invalidates interpretation of the data, leading to incorrect conclusions and potentially invalid improvement recommendations.

The first area of concern in choosing the appropriate target population, especially in active surveying, is the *working population*. This is the population from which the samples will be taken. For example, in a survey to measure the likelihood that purchasing agents will use an internet-based system, the target population of purchasing agents needs to be further defined. Are the respondent names going to be obtained from a professional association list or are companies going to be called at random with a request to speak with the purchasing manager? Are both purchasing agents and managers part of the working population? These issues need to be thought out and incorporated as part of the sampling plan. The concepts of bias and randomness, described later, are important issues to consider when defining the working population.

The second area of concern is the *timeframe* from which historical data are collected. *Historical data* come from actions that have occurred in the past, whether 100 years ago or the previous day. Collecting historical data, sometimes called passive data, is slightly different from collecting real-time process data, referred to as *active data*. When collecting historical data on a process, especially in a service environment, it is important to determine the timeframe from which you will gather samples. In the absence of common sense, you might decide that because the operation has been in place for the past 5 years, sample data should be gathered from the last 5 years. This would likely be an incorrect decision, because the operation has probably changed in the past 5 years. Remember, the point of collecting data is to understand or characterize the performance of the process: Is it meeting expectations? What are the gaps between current and expected? Where are the weakest links in the process? What are the vital few vs. the trivial many? It is only through understanding the performance of the current state of the process, especially where the problems exist, that we can develop improvement solutions.

A guideline for determining the timeframe from which to sample data is to determine how long the process has been operating in its current state. Recall that in developing a process map, it is important to document the current state of the process — not what happened last year, and not what we think the process should look like, but what it is. The same rules apply: collect only the data that provide information about the current state of the process. Experience suggests that, in most companies, given the dizzying rate of change in recent times, using data older than 1 year may not make sense. The exception would be if you are searching for long-term trends in the data; however, in most Six Sigma projects, the need to intimately understand the current process performance dictates use of relatively recent data for statistical testing.

Collecting active data on a process over time requires a different approach. We usually need to collect active data on a process because there are no historical data, such as for the wait time in drive-in banking windows, which is not typically tracked. In these situations, a data collection form or check sheet is used. The frequency with which the data are collected is important and will largely depend on the objective of the data collection. In all situations, however, the frequency must capture variations in the process to ensure the sample accurately reflects the underlying population. This can be problematic for processes with very long cycle times, such as product development; the team may need to get creative with estimates of past cycles coupled with ongoing measurements to arrive at a better understanding of the process. In addition to the issue of sampling frequency, if the process crosses multiple shifts or locations, samples must be taken from each to ensure the sample data represent the entire underlying population.

7.2.6.2 Selecting a Sampling Scheme

The sampling scheme describes in detail what data will be collected and how. The two areas of concern in selecting the sampling scheme are random sampling error and systematic sampling error.

Random sampling error is the difference between the sample measurements and the population measurements and is expressed as E, the measurement error. Intuitively, we know that as sample size increases, measurement error will decrease. We can also state that, given a constant sample size, measurement error increases as variability increases. This is discussed further in examining the required sample size below.

Unlike random sampling error, which is caused by chance variation in the process data, *systematic sampling error* results from poor design of the sampling plan. A good sample design, in which the sample data accurately reflect the target population, means that the units chosen as samples are free from bias and are randomly chosen.

Bias is when the sample measurements tend to deviate from the true population measurements, typically in one direction. A familiar issue in customer surveys, especially mail-based surveys, is *response bias*, or the tendency for only the very satisfied or dissatisfied customers to respond. Telephone surveys are less likely to involve response bias because names are chosen at random and their feedback is actively pursued. *Exclusion bias* results from excluding sample data from the entire population or process, such as not sampling across all shifts or locations. In addition, the individual collecting the data can consciously or unconsciously introduce bias by excluding certain data points and making errors in collecting or coding the data; this is called *operator bias.*

Randomly chosen samples are required to ensure true representation of the target population. Sometimes referred to as probability sampling, random sampling means that all units in the population have the same chance of selection. This is opposed to sample units selected on the basis of convenience or judgment, such as the first 10 occurrences of the day. Random number tables provide a random list of numbers to be sampled that is matched to an ordered list of the target population. Excel® and Minitab® can generate random number tables.

If the population of interest contains different groups or classes (strata) of data, stratified random sampling should be used to ensure randomness. Drawing samples independently from each group, with the size of the samples proportional to the group size, ensures that the sample measurements are truly representative of the population. Each group's sample units should be drawn in a random manner using random number tables.

In developing the data collection plan and the related sampling design, it is important to consider the possibility of stratified data. Otherwise, the sample data may exhibit unexplained variations that render questionable results, such as over-representation of a certain group's characteristic relative to the total. For example, in collecting sample data on turnover rates of nurses in large hospitals, the sampling design should consider stratification by geographical area. Given the regional nature of the economy and the number of hospitals, one region of the country may have significantly more turnover than another region. If, for example, there are more RNs in the Eastern U.S. but their turnover rates are lower due to economic reasons, the data need to reflect that. Without proper stratification of the data and subsequent understanding of root causes of turnover by geographical region, the improvement solutions may fail to adequately address the problem, as illustrated in Figure 7.10. Stratifying the turnover rates of RNs by geographical area reveals a significant difference between the two which will probably lead to different types of analysis and solutions for each of the areas.

7.2.6.3 Determining the Sample Size

Determining the appropriate sample size is perhaps the major source of concern in sampling design. I have received countless inquiries over the years along the lines of, "How many samples do I need?" The answer is always the same: "It depends." Many situational variables need to be considered in determining sample size, referred to as n, or the number of observations. The first is what kind of data we are sampling: continuous or discrete. The data type will determine the equation utilized to calculate sample size. The second variable is the objective of the study, which will determine the amount of measurement error we are willing to accept and the confidence we have in the final estimate.

Recall from the discussion on random sampling errors that as sample size increases, measurement error decreases. Also, as the sample size increases, the

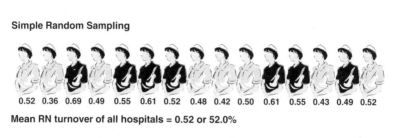

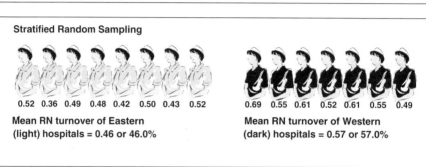

FIGURE 7.10 Simple random sampling vs. stratified random sampling; light gray is Eastern, dark gray is Western.

confidence we have in the sample measurements increases. Finally, as the inherent variability in the process data increases, the sample size needs to be larger, given a constant measurement error and confidence level. So, we have three variables to deal with in determining sample size for a particular data type: (1) variability of the population data, (2) measurement error, and (3) confidence level. Sample size equations sound something like this: "To ensure that the sample mean is within ±5% (measurement error) of the true population mean, with 95% confidence (confidence level), 383 samples should be collected from a population of more than 5,000."

The measure of the variability of the population data is the population standard deviation, or sigma (σ). Intuitively, we understand that the more homogeneous the data, the fewer samples required to estimate the population measurements from the sample data. In calculating the sample mean, we must estimate the population standard deviation. The most accurate estimate comes from collecting a small sample of data points in a procedure called *sequential sampling*. The first estimate is used to calculate the first *n* (sample size), which may be refined as more data are collected and a better estimate of the standard deviation obtained. A second method to estimate standard deviation is to use the rule of thumb of *one sixth the range of the population data*.[4] If the range of data for RNs' years of work experience is 20, then the estimate of the population standard deviation is 3.33.

The second piece of information needed to calculate sample size is *E*, the measurement error. Measurement error is calculated as an interval, referred to as the *confidence interval*. Do not confuse this with the confidence level, which is described below. The confidence interval measures the precision of the estimate,

TABLE 7.4

Effect of Different Confidence Intervals (Measurement Error) on Required Sample Sizes at a 95% Confidence Level for a Population of 100,000

Acceptable Measurement Error (E)	Required Sample Size
±7.0%	196
±6.0%	266
±5.0%	383
±4.0%	597
±3.0%	1056
±2.0%	2345
±1.0%	8763

such as ±5% of the true measurement. The actual percentage to use when determining sample size is a judgment call by the Black Belt in conjunction with the project team and Champion, based on the objectives of the project. It is worth repeating that it is very important to understand and document the objective of collecting the sample data as one of the first items in the data collection plan. If, for example, the project involves estimating the average number of errors involved in delivering medicines to hospitalized patients to make a recommendation on whether a new delivery system is needed, a small range of error is probably in order. The cost of the error is too great to allow a large acceptable error range. On the other hand, in estimating the percentage of customers who are very satisfied with a particular service process, there may be more room for error, especially considering the relatively subjective nature of satisfaction surveys. The measurement error is usually between 3 and 9%, with 5% a very common measurement in service processes. Changing the acceptable confidence interval by 1% significantly changes the sample size, as demonstrated in Table 7.4.

The third variable in determining sample size is the *confidence level*, or α. The confidence level is a percentage stating the long-run probability that the sample measure is accurate. In estimating mean values, a confidence *level* of 95% means that 95 times out of 100 times, the sample mean will be within the stated confidence *interval* (e.g., ±5%) of the true population mean. A confidence level of 95% is typical in Six Sigma projects. Increasing the confidence level to 99% almost doubles the required sample size, as illustrated in Table 7.5.

TABLE 7.5

Effect of Increasing the Confidence Level from 95 to 99% with a 5% Measurement Error and a Population of 10,000

Confidence Level (α)	Required Sample Size
95%	383
99%	661

7.2.6.3.1 Determining Sample Size for Continuous Data
Recall that continuous data involve data points along a continuum, such as height, weight, and dollars — typically, data that we discuss in amounts. Continuous data is preferred over discrete if there is a choice. The formula for determining sample sizes for continuous data is:

$$n = \left(\frac{ZS}{E}\right)^2 \tag{7.4}$$

where Z = the standardized Z score at a specified confidence interval (e.g., a 95% confidence level = 1.96 Z score); S = sample standard deviation, or an estimate of the population standard deviation; E = acceptable measurement error, expressed as a number.

For example, imagine that a Six Sigma team wants to estimate the sample size required to determine the mean time to process an insurance claim with a 95% confidence interval. The standard deviation is estimated at 29 min. The acceptable measurement error (E) is ±4 min from the mean. The calculation is:

$$n = \left(\frac{ZS}{E}\right)^2 = \left[\left(\frac{(1.96)(29.0)}{4.0}\right)^2 = \left(\frac{56.8}{4.0}\right)^2 = (14.2)^2\right] = 202 \tag{7.5}$$

7.2.6.3.2 Determining Sample Size for Discrete Data
When data are organized in discrete units, such as proportion of defective billing statements, a different formula is used to calculate sample size. An initial estimate of the proportion is required to use the formula. It is best to take a small sample (10 to 20 observations) prior to using the formula in order to initially estimate the proportion defective. The formula is:

$$n = \left[\frac{(Z^2) \cdot pq}{E^2}\right] \tag{7.6}$$

where Z^2 = square of the standardized Z score at a specified confidence interval [e.g., a 95% confidence level = (1.96)²]; p = estimated proportion of correct observations (successes); $q = (p - 1)$ = estimated proportion of defects (failures); and E^2 = square of the acceptable measurement error.

For example, if a Black Belt wants to determine the sample size required to estimate the proportion of defective billing statements (one or more errors on a statement) with 95% confidence and with a measurement error of no more than ±5% between the true proportion and the sample proportion, a sample size of 227 is required, as illustrated in Equation 7.7. The Black Belt estimates the proportion of defective billing statements at 18%, based on preliminary sampling:

$$n = \frac{(1.96^2) \cdot (0.82) \cdot (0.18)}{(.05)^2} = \left(\frac{0.567}{0.0025}\right) = 227 \tag{7.7}$$

7.3   DATA COLLECTION PLAN

Careful planning early in the project results in collecting representative and accurate data to measure baseline performance and sigma levels of the process, a primary deliverable of the Measure Phase. Representative and accurate data are also necessary to provide the team with a sound understanding of the current process, another key deliverable of this phase. If the appropriate key measures of the process are correctly identified at the start of the Measure Phase, developing the data collection plan should be relatively painless.

This is not to say, of course, that actually collecting the data is painless; in fact, it is probably one of the most time-consuming and difficult tasks of a Six Sigma project. Data collection is usually a case of feast or famine. Either there is an overwhelming amount of data available that must be sifted through to get to the relevant information or there are no data readily available that can be trusted to be accurate. I have witnessed many Six Sigma teams spend an inordinate amount of time and frustration on this activity. Sometimes having optimal data must be sacrificed in order to move the project along. The project Black Belt is the one to make that call because he or she is usually in the position to identify a proxy, or substitute, measurement. A typical example is gathering VOC data from all customer groups identified in the S-I-P-O-C. Although it is desirable to gather firsthand knowledge of customer satisfaction levels through formal surveys, the time and expense often do not justify the results. We discussed in Chapter 2 that a reasonable proxy for customer satisfaction are customer complaint data. Collection, categorization, and analysis of this data by a trained eye can produce a useful measurement of satisfaction levels at a fraction of the time and cost. That said, it is important to recognize that complaint data are biased, in that only customers with a problem complain. Most Six Sigma projects, however, focus on improving the problems, so it is often valid to use complaints as a VOC tool.

7.3.1   GETTING STARTED

Regardless of the type of project — cycle time reduction or general improvement — a detailed process map is usually a good first step in the data collection effort. A detailed process map, described further in Section 7.5.1, assists in understanding where the defects are in the current process, such as long delays because of hand-offs or approvals. It also helps identify discrete cycle time steps in the process, which leads to understanding how cycle time data need to be collected. For example, mapping the process of responding to customers' written inquiries highlights particular points in time that need to be measured. These include when the correspondence is received, when it is logged into the computer, when it is sent out to an external party for additional data, when it is received back in-house, and when the customer's answer is sent out and the case closed. Identifying these critical time milestones through process mapping leads to a sound data collection plan for capturing the data.

A critical component in the data collection planning process is designing data collection forms. A check sheet (Figure 7.11) is the primary tool for collecting active data over time or existing data through manual sampling. If you are collecting

Check Sheet for Collecting Defect Data on Outbound Letters Collector Name _____ Today's Date _____ Dates in which letters were sent _____ Instructions: Place a tally mark for each defect occurrence next to the reason. If there are two defects on one letter under the same category, count as two separate defects, e.g., if there are two spelling errors on one letter, tally two defects under spelling		
Type of Defect	**Tally**	**Subtotals**
Spelling	⦀⦀ ⦀⦀ ⦀	12
Grammar	⦀⦀ ⦀⦀ ⦀⦀ ⦀⦀	19
Text not left justified	⦀⦀	4
Times roman 12-pt font not used	⦀⦀	3
Incorrect customer name and address	⦀⦀ ⦀⦀ ⦀⦀ ⦀⦀ ⦀⦀ ⦀⦀	29
Incorrect adjustment amount	⦀⦀ ⦀⦀ ⦀⦀ ⦀⦀ ⦀⦀ ⦀⦀ ⦀⦀ ⦀	38
TOTAL DEFECTS		105

FIGURE 7.11 Example of a check sheet to collect data.

existing data through database extracts, you need to give some thought to the extract procedures and the spreadsheet design. Otherwise, you may waste considerable effort either collecting data that are not useful or failing to collect a data element that is needed.

When collecting active data on a process using a check sheet, consider the person who may be collecting the data. It may be unreasonable to ask clerical workers who perform the process to collect data if their jobs could be eliminated as a result of the project. Not only is there a good probability of bias in the data, it is simply not fair to ask employees, especially those not on the Six Sigma team, to participate in a process to eliminate their job. The exception may be where job guarantees are in place. Instead, collect recent existing data if possible using an objective person. Not only will it likely save time and effort, but the integrity of the data may be higher due to absence of bias.

Check sheets to collect the data should be simple and self-explanatory. The concept of *operational definition* plays an important part in designing the data collection form. The operational definition is a definition of the term that is specific and leaves little room for misinterpretation.[5] Instructions and data elements on a check sheet need to be operationally defined to ensure data integrity. Words mean different things to different people. If, for example, data need to be collected on the number of defects found on outgoing customer letters, the term defect needs to be further defined. Do defects include just spelling errors? How about grammatical errors? What about the centering of the text on the page? Unless clear operational definitions are used in the data collection form, the resulting data are probably suspect, and suspect data are a major problem in a Six Sigma project. Because the Six Sigma approach emphasizes data analysis and measurement, incorrect data can lead to incorrect conclusions and solutions that do not address the true root causes. In this case, everyone loses. The importance of spending the appropriate time and effort early in the project to collect accurate data that represent the true voice of the process and the customer cannot be overstated. One way to ensure appropriate and accurate data is perfectly clear instructions through the use of operational definitions.

7.3.2 BEST PRACTICES AND BENCHMARKING

A critical component that is often overlooked in planning for data collection is collecting data on industry best practices and benchmarking. *Best practices* refers to the practices and strategies used by top-ranked companies in the particular area of interest. *Benchmarking* refers to the process of gathering metrics for a particular area of interest in order to understand how your company measures up against others. *Benchmarks* are the actual metrics used for the comparison. The three are interrelated in that most benchmark data are gathered through benchmarking from best practices companies. Benchmarking involves identifying the process or function under improvement consideration, conducting research to identify best practices, obtaining process performance data from the best-in-class companies for that particular process, and measuring the gap between your company and the best-in-class company's performance. Figure 7.12 provides an overview of the benchmarking process.

Benchmarking as a quality improvement tool became popular in the 1990s, after Robert C. Camp of the Xerox Corporation published the bestselling, *Benchmarking: The Search for Industry Best Practices that Lead to Superior Performance.* Xerox was able to slow their market share loss to Japanese copier companies through understanding the best practices of not just other copier companies but companies across all industries who excelled in a particular process. For example, Xerox benchmarked their logistical functions against those of L.L. Bean, the mail order clothing catalog, once they determined that L.L. Bean had best-in-class logistical processes. During the benchmarking process, Xerox discovered why Bean was superior in this area and sought to emulate some of their practices.[6]

Although the popularization of benchmarking is attributed to Xerox, one of the earliest documented benchmarking activities involved the leaders of the young Toyota Motor Company visiting Ford during the 1930s and '40s. Based on the information gathered during many benchmarking visits, Toyota was able to understand the practices of the largest auto manufacturer in the world and applied their research to their own company.

The internet has made the process of benchmarking significantly easier than in previous years. Best practice research can now be done in a matter of hours, and the amount of benchmarking data for a wide variety of processes and functions available at no cost is truly amazing. Benchmarking groups dedicated to a particular industry, such as the call center industry, have also become common in recent years.

7.4 CHOOSING STATISTICAL SOFTWARE

A Six Sigma Black Belt needs access to a statistical software package that goes beyond the very good but basic Excel statistical functions. Many software packages are available, and often the decision of which one to use is the one that the company is already using. SAS® is a widespread application that provides a modular approach to data analysis, including robust statistical analysis functionality (SAS is an acronym for statistical analysis system). The SAS Institute in Cary, North Carolina, provides excellent public courses as well as on-site training. In recent years, it has attempted to capture the less technical market by providing a menu-driven, Windows®-based

---

PLANNING

1. **Identify what is to be benchmarked:** Individuals need to identify the processes they perform. They also need to determine what statistics are available both internally and externally. External sources may be from industry or trade associations, publications, governmental sources and library material.

2. **Identify comparative companies:** Identify companies, competitors or not, that have superior business processes. Benchmarking must be conducted against leadership companies and business functions regardless of their industry.

3. **Determine data collection method and collect data:** Understand which statistics are relevant, timely and accessible. Refine questions in detail. Contact other firms willing to serve as benchmark partners. Perform site visits. Realize that benchmarking is a process to collect not only quantifiable metrics but also industry best practices.

ANALYSIS

4. **Determine current performance "gap":** Identify the areas of opportunity for comparative analysis. This may cross industries and functional areas.

5. **Project future performance levels:** If changes are to be made, where will these measures of performance be in one year? Two years? Ten years? Understand that benchmarking is an ongoing process where today's performance is constantly assessed against future leaders.

INTEGRATION

6. **Communicate benchmarking findings and gain acceptance:** Discuss with management the gaps that have been determined and what increases in levels are possible.

7. **Establish functional goals:** Based upon discussions with management, establish a plan of action. Fine tuning of goals may be needed.

ACTION

8. **Develop action plans:** In detail identify what changes will be made to a specific process. This will include technology changes, work-flow revisions, reduction of non-value added steps, and similar actions. Define measurement criteria for these changes.

9. **Implement specific actions and monitor performance:** Implement the action plans and measure performance against it.

10. **Recalibrate benchmarks:** Review the benchmarks that have been established to determine if targets remain relevant and achievable. Identify new measurements if necessary.

MATURITY

- Leadership position attained
- Practices fully integrated into processes

FIGURE 7.12 Benchmarking process. (From Camp, Robert C., *Benchmarking: The Search for Industry Best Practices that Lead to Superior Performance*, ASQ Quality Press, Milwaukee, WI, 1989. With permission.)

application in addition to its signature programming-oriented product. On the other hand, Minitab, Inc. (State College, PA) appears to be emerging as the standard for Six Sigma projects. Although Minitab provides advanced statistical techniques that many Black Belts will never use, its ease of use and widespread adoption among Black Belts provide significant advantages. Both of these software packages, along with a relative newcomer, Statgraphics® (Manugistics, Inc., Rockville, MD), are described below.

7.4.1 MINITAB

Minitab was originally developed to teach statistics at Pennsylvania State University in 1972. One of the three founders of the package and later the company, is Dr. Brian Joiner, a renowned TQM expert and disciple of Dr. Deming. The founders' objective was to provide an easy-to-use package so that students could concentrate on learning the complexities of statistics rather than the complexities of computing.[7] It has advanced through the years to become almost completely menu driven, with a data entry screen similar to Excel (i.e., columns of data specified as variables). A programming window offers additional functionality beyond the menu-driven commands. It is fully compatible with Excel, facilitating data transfer between the two programs in one easy step, and it uses standard Windows interfaces. Although originally developed for academic use, nonacademic users, primarily businesses, now account for more than 70% of the users. I use the latest release, Minitab 13, in my Six Sigma project analyses as well as to teach Black Belt and Green Belt classes. Other training and consulting firms, such as the Six Sigma Academy, the Juran Institute, the ASQ, and PriceWaterhouse Coopers, use and recommend Minitab in their Six Sigma training.[7]

7.4.2 SAS JMP®

SAS was founded in 1976, primarily for application in pharmaceuticals.[8] SAS offers integrated software through a modular approach; at the time of this writing, 46 separate modules are available. The JMP (pronounced "JUMP") module is the most practical in Six Sigma applications because it joins statistical analysis with graphics capability, similar to Minitab and Statgraphics. Although menu driven through a Windows interface, JMP also allows the user to customize data exchange and reporting using its own scripting language.

As with most SAS modules, the level of statistical knowledge required to fully utilize the software is on the high side relative to Minitab and Statgraphics.

7.4.3 STATGRAPHICS

The relative newcomer to the statistical software world is Statgraphics, developed in the early 1980s. Version 5, Statgraphics PLUS®, provides the same menu-driven Windows interface of Minitab and SAS JMP. Also similar to Minitab 13, Version 5 is specifically geared toward the type of analysis required for Six Sigma projects and appears to have been developed with an eye toward Six Sigma application.[9] It even uses the nomenclature of Six Sigma, such as reporting DPU (defects per unit) and DPMO (defects per million opportunities), within its process capability functions. Although Minitab provides the same measurements, it does not specifically refer to them as DPU and DPMO, but rather in terms of benchmarking Z and parts per million.

Because of its emphasis on Six Sigma, Statgraphics PLUS is likely to give Minitab some hefty competition in this particular market segment.

7.5 MEASURE TOOLS

Whereas Define Phase tools help to define the project and understand the VOC, the tools included in the Measure Phase are intended to more clearly understand the voice of the process. Measure Phase tools include process maps, Pareto charts, C&E diagrams, 5 Why analysis, histograms, and control charts.

7.5.1 PROCESS MAPS

Rarely is a Six Sigma project successfully completed without a detailed process map. Individuals who have been in the process improvement field for awhile process flow almost any concept as second nature. As covered in Chapter 3, all work is a process, with inputs, process steps, and outputs. Most individuals learn what they can visualize, and the process map is a wonderful tool to take a list of process tasks and bring it to life in a meaningful way. Process maps provide a common level of knowledge among many diverse individuals similar to what operational definitions do for measurement terms.

There are two basic types of process maps, each with a different purpose. The first is a process flow diagram, in which the sequence of steps in a process is graphically represented through special purpose symbols, as illustrated in Figure 7.13. The second is a process deployment map, also known as a cross-functional map, in which the graphical symbols are used in a matrix arrangement to show how the process flows among several areas, people, or departments.

Guidelines for developing detailed maps that are easy to understand include:

- Involve the employees who actually perform the process. It can be quite interesting to watch those who "manage" the process participate in a process mapping session with those who perform the process steps. Typically the manager will begin describing the process in an ideal state — the way it is supposed to happen. An effective meeting facilitator, usually the Black Belt or Green Belt, draws out the participation of the process performers by asking questions such as, "Is this the way it happens on a routine basis?" With a little encouragement, the individuals who perform the process start to describe how the process really works and it becomes apparent that "exceptions" to the process are more likely to be routine. Following this path a little longer leads everyone to a greater understanding of just why the process takes three times as long as it should.
- Map the "as is" process rather than the "should be" or the "to be." It is often difficult to focus on the way the process actually works rather than how participants expect it (should be) or want it (to be) to work. This is especially the case in an improvement project because the team is already thinking ahead to improvements. The facilitator must keep everyone focused on the current process. In the Improve Phase, the opposite scenario often occurs: participants want to focus on the "as is" process rather than breaking the paradigm and developing truly creative and innovative "to-be" process solutions.

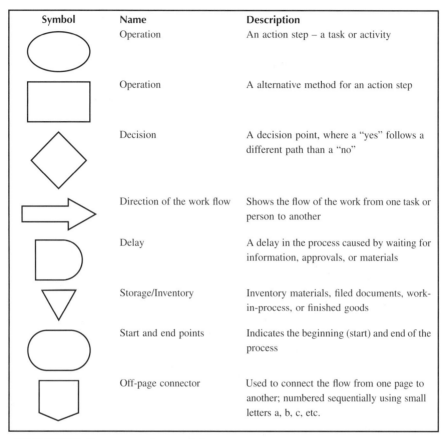

Symbol	Name	Description
	Operation	An action step – a task or activity
	Operation	A alternative method for an action step
	Decision	A decision point, where a "yes" follows a different path than a "no"
	Direction of the work flow	Shows the flow of the work from one task or person to another
	Delay	A delay in the process caused by waiting for information, approvals, or materials
	Storage/Inventory	Inventory materials, filed documents, work-in-process, or finished goods
	Start and end points	Indicates the beginning (start) and end of the process
	Off-page connector	Used to connect the flow from one page to another; numbered sequentially using small letters a, b, c, etc.

FIGURE 7.13 Process mapping symbols.

- Keep participants focused on the task at hand and avoid letting the group go off on tangents such as how we "used to do it." Although some of this exchange is valuable, the tendency to describe every possible situation that happens in a process is not a good use of anyone's time. This is especially difficult to control because those who are closest to a broken process feel a lot of frustration on a day-to-day basis and this may be the only opportunity to vent. In some cases the 80/20 rule may need to be invoked — document the 20% of the situations that result in 80% of the problems.
- Use colored Post-It notes on a flip chart to record the process steps; documentation with Visio® or some other software can be done later. Post-It notes save a lot of time and effort because the first run-through of the process is likely to be messy, with a lot of print and paper ending up in the garbage.
- Agree at the beginning of the session the boundaries of the process — where it begins and ends. The scope of the project as described in the project charter may need to be referred to for clarification on the boundaries.
- Start each step with an action verb, such as "Access file" or "Pick up mail." Keep process step descriptions short and at the same level of detail

throughout the map. If a detailed description is required on one or two steps, footnote the detail so the main map can flow smoothly at the same level of granularity.

- Do not include the names of employees who perform the work unless it is absolutely necessary to the understanding of the process. The idea is to focus on the process rather than who performs it. It may help to think of a widget moving through the process, whether the widget is an order, a phone call, or a claims form. The widget method provides everyone in the session the same perspective and facilitates clearer documentation of the process.

- Clearly indicate inputs and outputs of process steps and highlight what directly touches the "customer" of the process. In the Analyze and Improve Phases, these steps need to be examined more closely in order to optimize the "moments of truth." Moments of truth are those critical opportunities in which the customer has direct contact with the company and the process. What happens in these critical moments determines whether the customer walks away delighted, satisfied, or dissatisfied with the company's service. It is especially critical in service industries, where absence of a tangible product means that the company is judged primarily by the quality of service rendered. Recall from Chapter 4 that, in the mind of the customer, the person answering the phone IS the company. Customers do not know the company's internal structure and initiatives, how many agents are staffing the phones, and often even where the company's call center is located. Whether the customer leaves with a positive or negative experience depends to a large degree on the short transaction between agent and customer: the moment of truth. Thus, the process map needs to highlight these critical encounters so that the Six Sigma team can focus on these steps.

- When the process reaches a decision point or branches off in one or more directions, follow one branch until it either ends or goes back into the mainstream process.

Remember, these are *guidelines* in developing effective process maps; every situation is different and may require departure from the guidelines. The idea is to make the tool work for the team in the manner most effective for process redesign. After the process mapping session is over, one or two team members need to be assigned the task of translating the Post-It notes on the flip chart to a Visio® or other flow charting file. Once this is completed, the file should be distributed to all who participated in the mapping session for review and comments, and a follow-up meeting should be held to review a printed copy and make necessary changes as a group.

7.5.2 Pareto Charts

After a detailed process map has been completed, the next step is separate the vital few from the trivial many by developing a Pareto chart of the data. The chart is named after an Italian economist named Vilfredo Pareto, who in 1897 pointed out that 80% of wealth was owned by 20% of the people. The 80/20 rule was used primarily in economic analysis until 1934, when Dr. J. M. Juran applied the concept to the field of quality control, noting that most defects are the result of relatively few causes.[10] Dr. Juran named his defects analysis after Pareto, and thus Pareto charts have become a familiar and useful tool in the process improvement toolbox.

Pareto charts are the graph of the data and Pareto analysis is the interpretation of the charts. They are used at this stage of the project to understand where the Six Sigma team should focus its efforts. The critical process output measures, or Y's, have already been identified through VOC and other data. These measures usually relate to accuracy, timeliness, responsiveness, and access issues. Input and process measures, which usually relate to internal efficiencies, should have also been identified. Pareto charts help to identify the critical X's that have the most impact on the process: What are the few defects (inputs) causing most of the problems (outputs)? We are usually interested in classifying defect reasons by their frequency of occurrence, leading the team to focus their efforts. However, Pareto charts are useful in areas other than defect analysis, such as identifying the vital sources of revenues from the trivial many sources.

Figure 7.14 provides a Pareto chart of the frequency of the defects identified in the check sheet example (Figure 7.11). It is easy to see that customer information and adjustment amount account for the majority (63.8%) of the problems even though they represent only 33% of the defect types. Based on the Pareto chart, where should the team initially focus their efforts?

This example illustrates the usefulness of Pareto analysis early in the Measure Phase. Through systematic measure and identification of the vital few, the team can now focus on the small number of areas that have the greatest potential to effect overall improvement of the process.

Before constructing a Pareto chart, you must carefully consider what you want to analyze. The most common types of Pareto charts in Six Sigma relate to defects and sources of defects. Of course, the team may want to analyze other items of interest through Pareto charts, such as seasonal effects (e.g., What months have the greatest volume of inquiries?). The following example of construction of a Pareto chart focuses on defects analysis (see Figure 7.14). Here are the steps in manually developing the chart:

1. Operationally define and categorize the defects by type, such as spelling errors, wrong address, incorrect formatting, and wrong font size.
2. If defect data have not been collected, decide on the target population and whether sampling will be used to collect the defect data. Develop the sampling plan if necessary and collect the data by defect type according to the plan.
3. Create a three-column table with the following headings: Type of Defect, Frequency, and Percent of Total (Table 7.6). List the defect descriptions in the Defect column and their frequencies in the Frequency column. Divide the frequencies for each type of defect by the total defects and enter that in the Percent of Total column.

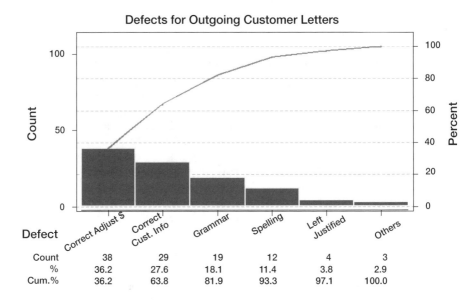

FIGURE 7.14 Pareto chart of defects for outgoing customer letters.

4. Draw one horizontal and two vertical axes on the Pareto chart. The horizontal axis shows the defect types in descending order of frequency from left to right. The left vertical axis indicates the frequency for each of the defect types. The right vertical axis represents the cumulative percentage of defect frequencies.

5. Plot the bars on the Pareto chart. Using a bar graph format, draw the corresponding bars in decreasing height from left to right using the frequency scale on the left vertical axis. To plot the cumulative percentage line, place a dot above each bar at a height corresponding to the scale on the right vertical axis. Then connect these dots from left to right, ending with the 100% point at the top of the right vertical axis.

A statistical software package, such as Minitab, should be used to develop the Pareto chart. Otherwise, you may spend more time developing the chart than analyzing it.

7.5.3 CAUSE-AND-EFFECT DIAGRAMS

In searching for the critical X's that cause the critical Y's, one very useful tool is the C&E diagram — the X's are the causes and the Y's are the effects. C&E diagrams are also referred to as fishbone diagrams and Ishikawa diagrams (Figure 7.15). In 1953, Professor Kaoru Ishikawa of the University of Tokyo first used a C&E diagram to summarize the opinions of engineers relating to a manufacturing quality issue. The technique proved so effective it was included in the Japanese Industrial Standards terminology of Quality Control. They define a C&E diagram as "a diagram which shows the relation between a quality characteristic and factors."[11]

TABLE 7.6
Data for Constructing a Pareto Chart of Customer
Correspondence Defects

Type of Defect	Frequency	Percent of Total
Spelling error	12	11.4%
Grammar error	19	18.6%
Text not left justified	4	3.8%
Times roman 12-pt font not used	3	2.9%
Incorrect customer name and address	29	27.6%
Incorrect adjustment amount	_38_	_36.2%_
TOTAL DEFECTS	105	100%

Typical Six Sigma projects involve a systematic process for *identifying relationships among quality characteristics and the factors* that contribute those characteristics. The C&E diagram helps to organize the many causes or factors (critical X's) that contribute to the outcome (critical Y), providing structure to a seemingly vague and messy situation. It is helpful to use the tool at the beginning of the Measure Phase to help organize the team's ideas and start to identify critical X's, with the outcome as the project problem statement. Further along in the project, once the team has identified the project's critical Y's, a C&E diagram should be generated for each critical Y. C&E diagrams can be generated at various levels of detail: at the project level with the project problem statement as the outcome, at the macro level with a critical Y as the outcome, and at the micro level with a cause that was identified in the macro level diagram as the outcome. Further "drilling down" to the micro level may be necessary to help the team move past symptoms to true root causes.

Generating a useful C&E diagram is difficult but generally worth the time and effort. It is a team development effort and involves collecting the "wisdom of the organization."[12] In addition to the Six Sigma team members, subject matter experts should be included to identify a comprehensive list of causes. C&E diagrams are similar to process mapping tools. The usefulness of these tools for solving problems is directly correlated to the amount of effort involved in making them. Although seemingly simple in concept, properly developed diagrams and process maps provide each team member with the same level of understanding of the problem to be solved, supply details on the nature of the problem, and contribute solutions. Often, Six Sigma teams are so fact driven (which they should be) that the experience of those knowledgeable of the process is discounted as being "subjective." This is a mistake — the collective organizational wisdom combined with objective data analysis result in the best solutions.

7.5.3.1 Constructing Cause-and-Effect Diagrams

Making the C&E diagram is a group effort; so it is helpful to use a flip chart and Post-It notes. The categories of causes, or "bones" of the fish, are chosen to fit the particular situation. In manufacturing, a common practice is to categorize by function,

Cause and Effect Diagram for Service Industries

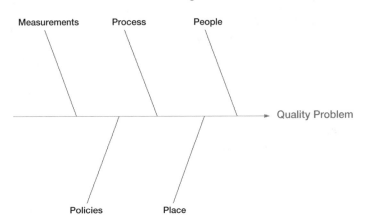

FIGURE 7.15 Cause-and-effect diagram template for service industries.

using the "five M's": man, methods, machines, measurement, and materials. In services, these are replaced with the "four P's": people, place, process, and policies. A measurements category is usually added to this type of diagram (Figure 7.15).

The categories are flexible and should be chosen to the specific purpose. For example, if late deliveries are the quality characteristic of interest, the C&E diagram could have just two bones — vendor and client — with subbones off each.

The steps in making a C&E diagram are:

1. Choose the quality characteristic or outcome to be studied and write it at the end of a horizontal line. Only one outcome is diagrammed at a time.
2. Choose the categories that contribute to the outcome. I highly recommend that a team with little experience in developing C&E diagrams start with the four P's and M. Otherwise, the team may spend a lot of time discussing and debating the proper categories. Once the initial diagram is created, subsequent diagrams can be made with special categories if necessary.
3. Brainstorm with the team to identify secondary causes to the primary causes (the four P's and M). There are likely tertiary causes to the secondary causes, which will need to be documented with further branching.
4. After listing all causes, review the diagram as a whole to make sure all causes are identified, eliminate duplicate causes, or change categories of causes if necessary (e.g., move poor training from people to process).
5. Assign an importance weighting to each factor: 1 (significant contributor to the effect), 3 (strong contributor), 6 (medium contributor), and 9 (moderate contributor).
6. List the factors in order of importance.

The next step in the process is to analyze the diagram and make actionable plans to further measure or research causes, starting in order of importance. It is probably worthwhile to have a second team session to review and analyze the completed diagram after some time has passed. The following questions should be asked in analyzing the diagram:

- Has this factor been documented as a problem before? If so, the documentation should be reviewed to make sure the team is not duplicating efforts or moving toward a solution that was previously rejected.
- Are there existing performance measurements for the factor? As an example, one of the factors contributing to repeat calls may be system downtime, which is probably already measured. Through examination of historical system downtime measures, the team can assess the magnitude of this factor.
- Does this cause interact with another cause? Interaction between factors needs to be identified so the team will understand that fixing one factor may significantly change another factor — either positively or negatively.

The following are helpful hints for developing a C&E diagram:

1. Express the causes as concretely as possible so that further measurement of them can be made. Generalities and abstract concepts are not very useful for taking action on or pursuing further measurement of root causes.
2. Separate symptoms from true root causes when feasible.
3. Keep the team focused on the task at hand. It is easy to digress to defensive justification or rationale of a particular cause. Although it is important to get all the ideas on the table, the team may need frequent reminders of the purpose of the diagram — to identify root causes rather than justify them.

Ideally, the C&E diagram is used along with the Pareto chart to lead the team toward further identification of the critical X's that drive the outcomes and customer CTQs of the process. This helps narrow the focus to a manageable number of areas on which to focus, leading to a clearer understanding of the relationships between the process inputs and the customer CTQs. Confirmation of the root causes is the main deliverable of the Analyze Phase, leading to development of intelligent solutions in the Improve Phase.

7.5.3.2 The 5 Whys

Similar in purpose to the C&E diagram is the 5 Whys tool. The idea is that by asking "why?" five times, you can move beyond the symptoms of a problem to the true root cause. It is natural to stop at the first or second "why" and identify that as the root cause. However, continuing to ask why ultimately leads to true root causes. Five whys are usually sufficient to get to the bottom of the issue.

7.5.4 Histograms

After the Pareto analysis has helped the Six Sigma team focus their efforts by identifying the "vital few," a clearer understanding of the vital few elements is in order. If, for example, Pareto analysis indicates that claim type A is the source of

68% of all timeliness defects, even though it represents only 25% of claim types, the team knows that claim type A should be investigated further. All kinds of questions are likely to be posed, such as: What is the average and the distribution of claim type A cycle times? Does it vary by day of the week or by week of the month or by time of year?

Histograms are graphical representations of data distributions. The graphical representation helps visualize patterns in the data variation that are not readily apparent in tabular forms of the data, especially the *shape* of the data distribution. All data vary to some extent. The nature of the variation determines the shape or distribution of the data. As described in Section 7.2, a common distribution found in nature is the bell-shaped normal distribution, in which most data are clustered around a central point with decreasingly less data from the center on either side.

To reach a level of six sigma, or 3.4 DPMO, an extremely stable process must be in place. By stable, I mean that the variation of the process output can be predicted with a high level of confidence — this can be the case only if variation in the process itself, as well as the output, is relatively small. In the case of Six Sigma, this means we can predict with some level of statistical confidence that the output will vary no more than .000097% from its historical mean. Measurements of this type are derived through the use of *statistical process control* (SPC) charts, which are described in Section 7.5.5. For now, histograms will provide the first step toward characterizing the variation in a process or set of data points.

Figure 7.16 shows a histogram (created in Minitab) of the length of phone calls into the case study call center. The time to complete calls is classified into intervals on the X axis, and the frequency of occurrence of each interval is noted on the Y axis. The pattern is not the classic bell-shaped normal distribution of clustered data around a center point; rather, it appears that there are two smaller bell-shaped distributions. This is called a bimodal distribution and is common when two distinct types of processes are at work. Further stratification of the data reveals the longer call types (the second peak) are calls related almost exclusively to billing disputes (i.e., the customer disagrees with a charge on the billing statement). This requires that the call agent "set up a case," for which a significant amount of information relating to the disputed charge is entered into the computer and a case number is assigned to the caller for future reference. The shorter calls were primarily inquiries, such as: What is my credit limit? When was my last payment received? and similar-type questions that do not require the same amount of data gathering and entry as the dispute calls. Based on the histogram, the Six Sigma team chose to treat each of these call types separately, because improvement solutions were likely to address each distinct process.

7.5.4.1 Constructing Histograms

Constructing a histogram by hand is tedious and error prone and is therefore not recommended. Spreadsheet software, such as Excel, and special-purpose statistical software, such as Minitab, are readily available and easy to use for this purpose. The steps described below are for constructing a histogram using Excel and the bar

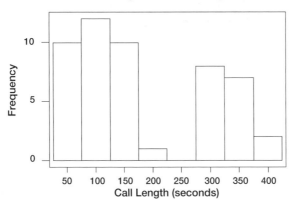

FIGURE 7.16 Histogram of the length of calls in the call center case study.

charting feature. Although Excel's data analysis tool pack possesses a histogram function, the manual method will be used here to provide a clearer understanding of how histograms are developed from raw data, using the sample data in Table 7.7. The data represent the number of seconds to close out a Card One customer billing dispute. The histogram is generated as follows:

1. Arrange the raw data in a column and identify the high and low values and the range between the two. From Table 7.7, the highest value is 800 sec and the lowest is 727 sec, for a range of 73.

2. Determine the number of class intervals. This is determined so that the range is divided into equal widths of 5 to 15 intervals, depending on the number of data points or observations. Use the table below as a rule of thumb:

Data Points	Number of Intervals
<50	5
51–200	8
201–500	10
>500	12–15

For the data in Table 7.7, the range is 73, so we will try 8 as the number of intervals. Note that intervals are also called "buckets."

3. Calculate the interval width by dividing the range by the number of intervals. If necessary, round this number to the nearest 5 or 10, depending on the data. In the example data, the calculation is 73/8 = 9.125, so we will round to 10. The intervals are then 725–734, 735–744, 745–754, ... 785–794, and 795–804.

TABLE 7.7
Data of Processing Time in Seconds

800	771	757	769	750
751	766	800	754	748
734	765	765	771	759
758	759	752	766	733
775	750	733	762	766
752	800	742	754	748
752	771	771	749	751
744	727	753	744	771
758	800	766	761	752
758	758	770	800	759

4. List the class intervals in a column and tally the number of observations in each interval; list the frequencies in a column next to the intervals. The lowest value (727) must be within the first interval and the highest value (800) must be within the last interval. Because we are using Excel to develop the histogram, we can sort the column of data and easily identify the frequencies in each bucket:

725–734	4
735–744	3
745–754	14
755–764	11
765–774	11
775–784	1
785–794	0
795–804	5

5. The final histogram, created through the bar charting feature of Excel, is shown in Figure 7.17.

7.5.4.2 Interpreting Histograms

The shape of the histogram provides insight into the pattern of variation in the data (Figure 7.18). One can then make hypotheses as to why the process data are exhibiting this type of variation. This helps the team begin its search for the root causes of the variation, which are validated further in root cause analysis. Distributions may take the following shapes:

Bell shape. This is the classic normal distribution, with the majority of the data clustered around the center and less data as you move away from the center. Shapes similar to this are the most common type of variation.
Bimodal shape. Sometimes called the twin peak shape, this distribution shows two distinct peaks, indicating two distinct processes at work here. The data

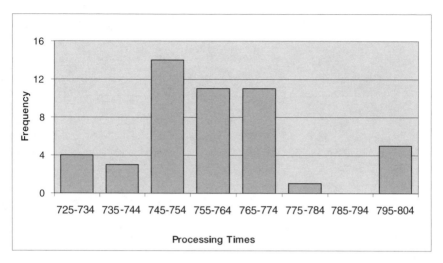

FIGURE 7.17 Histogram of processing times.

values in the first peak should be separated from those of the second peak and further analysis performed to determine if there really are two separate processes at work. Recall from Figures 7.16 and 7.17 on call length, we saw a bimodal shape because of the two distinct types of calls being handled.

Skewed shape. This can be either positive (skewed to the right) or negative (skewed to the left) and typically occurs when there is a lower (right skewed) or upper (left skewed) limit to the data. A common example is monthly payment due dates: the majority of the payments arrive in the few days prior to the due date rather than evenly throughout the month.

Truncated shape. This looks like one half of a normal bell-shaped distribution, with the other half cut off or truncated. This usually means that a screening process is in place to remove all output below (right truncated) or above (left truncated) a certain threshold value.

Plateau shape. This type of distribution indicates similar frequency values in each interval class. This could be because the intervals are too wide to appropriately distinguish between the true variation in the data or it could mean several different processes are at work, each with their own mean value and variation around the mean.

Comb shape. This distribution is so named because every other interval has a lower frequency. It indicates incorrect width of the intervals or errors in the data. The first step upon seeing a comb-shaped histogram is to examine the data for accuracy and then consider different interval lengths.

Isolated peak shape. This is typically a bell-shaped distribution with a second small peak. This pattern can indicate several situations, such as a second process at work, a process anomaly that occurs infrequently, or a measurement error.

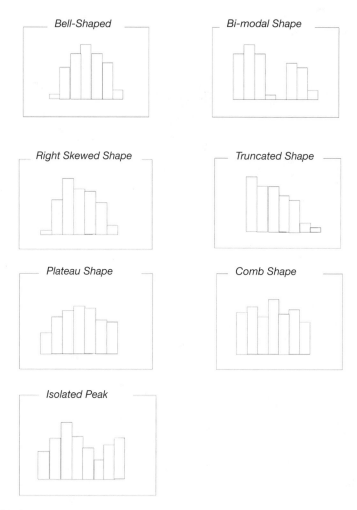

FIGURE 7.18 Types of histograms.

7.5.5 CONTROL CHARTS

The Six Sigma approach is based to a large degree on control charts. A process is in a state of statistical control when the outcome of a process can be predicted with some degree of certainty. The underlying concept is to compare what happened yesterday to what happens today and measure the difference. Variation in the process is measured historically over time and control limits are calculated based on the measurements. We then compare today's process variation with the historical variation, and if the difference is large enough, we take action.[13] It was not until the advent of mass production that the need for statistical quality control arose. Prior to Henry Ford's interchangeability of parts concept, quality was defined as an attribute intrinsic to an object, such as beauty or strength. In Ford's mass production world, repeatability and reproducibility became the new definition of quality.

Control charts have two basic purposes. The first is to analyze past data to determine past and current performance of a process. The second is to measure control of the process against standards. In the Measure Phase, control charts are used primarily for the first purpose. If a process is determined to be out of control in the Measure Phase, the process may need redesigning through a DFSS approach. It is common for DMAIC projects to become DFSS projects after initial measurements of process performance are obtained. Through the use of control charts, the degree of stability a process exhibits will assist in determining the appropriate approach. Control charts are also useful in the Measure Phase to determine if the current process is capable of meeting customer CTQs. Section 7.10 discusses process capability.

The second purpose of control charts, ongoing measurement against standards, is applicable in the Control Phase. Control charts at that point determine whether the intended result of a process change actually happens, particularly changes intended to have a positive effect on the process CTQ measures.

Dr. Walter Shewhart developed the concept of statistical process control charts in 1924 after identifying that variation in the manufacturing process led to a high number of nonconforming (defective) products. He developed a system to measure variation in the manufacturing process and to determine if the variation was such that nonconforming products were likely. Dr. Shewhart tested his system in the manufacture of telephones at Bell Labs with successful results, and he went on to develop an entirely new field of science called statistical quality control (SQC).[14] Statistical process control charts are the most widely used of the SQC techniques and are the only ones aimed at preventing manufacture of defective products by taking corrective action at the source of the problem. Other SQC techniques, such as acceptance sampling plans, are aimed at detecting defective products after they have already been produced. The preventive nature of control charts is what led Dr. Deming to become a prophet of statistical process control, using the technique to teach Japanese manufacturers that control of the process is the key to high-quality output.

Shewhart identified two types of process variation cause: chance and assignable. Chance causes are inherent to the process; there is variation in everything — it is just a matter of degrees. If the variation stays within selected limits, then the only variation is the chance day-to-day variation that is expected to occur because no two things are alike. Assignable causes occur when the variation in a process is outside the selected limits of chance cause. Something other than normal day-to-day variation is at work, signaling the need to take corrective action. Dr. Deming relabeled chance cause variation *common cause* and assignable cause variation *special cause*. As an example, consider driving to work each morning along the same route. Because of traffic lights or volume of traffic, there is small day-to-day variation in the time it takes to travel from home to work. This small variation is due to a common cause that is inherent to the process. One morning, there is an accident along the route that blocks traffic for 30 min. This is clearly a special cause situation, because it is not part of the routine process.

A process in which only common causes are present is said to be in a state of statistical control; that is, the process is stable and future output can be predicted

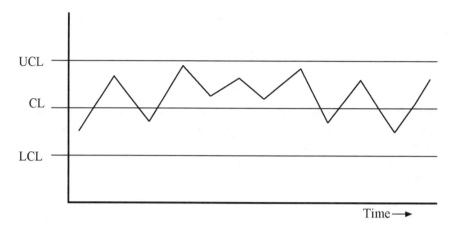

FIGURE 7.19 Example of a process in a state of statistical control. (UCL, upper control limit; CL, center line; LCL, lower control limit)

with a degree of statistical confidence (Figure 7.19). If special causes are present, the process is said to be out of control, which means that the process is unpredictable and unstable (Figure 7.20).

Recall from Section 7.2 that for normally distributed data, approximately 68% of the observations are within 1 standard deviation from the mean, 95% within 2 standard deviations, and 99.73% within 3 standard deviations from the mean. This means there is only a 0.27% chance of finding observations outside 3 standard deviations. In statistical process control, when an observation is detected outside of the 3 standard deviations limit, a special cause is likely at work; there is only a 0.27% chance that an observation will fall outside the 3 standard deviations limits if only common causes are present.

The basic construction of a control chart, as illustrated in Figures 7.19 and 7.20, is a center line and the upper and lower control limits. The center line is the average value of many data sets tracked over time and the upper and lower control limits are set to 3 standard deviations from the center line. Only about 0.27% of observations will go beyond the upper and lower control limits when the process is in a state of statistical control.

The first step in making a control chart is estimating the common cause variation. This is done through sampling small subgroups of 2 to 10 consecutive observations from an ongoing process and treating each subgroup as a separate data set. The subgroup is likely to consist of homogeneous data because the consecutive observations are comparatively close in time. The average value and range of dispersion of this subgroup data are a good estimate of the variation attributable to common causes. The subgroup averages and range are plotted over the time period. The most common types of control charts for continuous data are the X-bar and Range (R) charts. The X-bar chart records the averages of each subgroup and the R chart records the range within each subgroup. Recall from Section 7.2 that the range is one of two common dispersion statistics (along with the standard deviation) and is calculated by subtracting the lowest value from the highest value in a data set.

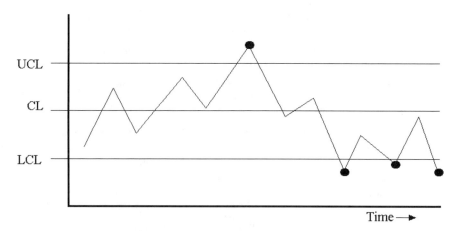

FIGURE 7.20 Example of a process that is out of control. (UCL, upper control limit; CL, center line; LCL, lower control limit)

In a state of statistical control, the individual observations *within* each subgroup vary, yet the average value of each subgroup is consistent from subgroup to subgroup; in other words, the variation *between* subgroups is random and due only to common causes. It is this comparison of the *within*-subgroup averages and ranges *between* subgroups that makes the control charts a very powerful, robust, and yet simple tool for measuring and controlling process variation. The R chart measures the degree of within-subgroup variation and the X-bar chart exhibits the variation between subgroups.

The unusual and unlikely occurrence of an event (0.27% chance) according to the natural laws of probability are why control charts are so effective. Not only are they valuable in determining when to take corrective action on a process, they are just as valuable for knowing *when not to take action*. Without a scientific basis on which to judge whether the variability in a process is within normal chance causes, the tendency is to overreact to normal day-to-day variation as though it was a special cause. This overreaction results in chaos and leads to an out-of-control process. Because the process is constantly being tampered with in reaction to common cause variation, there is no opportunity for the process to reach a state of control. The exact opposite of the intention of maintaining stability occurs. This is why control charts are the foundation for the "economic control of quality," which is part of the title of the landmark book in which Shewhart described his concepts and techniques.[15]

7.5.5.1 Types of Control Charts

Seven different control charts are typically used in transactional Six Sigma projects. Their use is dictated by the type of data (continuous or discrete), sample size, and type of defect measures. For continuous output variables:

1. X-bar and R charts: Used for continuous outputs such as time and weight. The most common type of continuous output measure in transactional services is duration of time, such as time to process an invoice or cycle

time of a purchase order. The X-bar represents the subgroup averages and R is the range within each subgroup. The X-bar chart tracks shifts in the process mean and the R chart tracks the amount of variability in the process.

2. X-bar and S charts: Similar to the X-bar and R charts except that the standard deviation (instead of the range) is used as the dispersion measure. When subgroup sizes are greater than 10, an S chart should be considered.

3. X and mR charts: When rational subgroups are not feasible, such as very long intervals between observations, X and mR charts are used. These charts are also called Individuals and Moving Range charts. Individual observations (subgroup size = 1) are the X's plotted over time and the moving range of the successive data points is used as the range measure to calculate control limits of X.

For discrete output variables (also called attribute or count variables):

1. nP charts: These charts are used for tracking the proportion of defective units when the sample size is constant. The output is classified as either defective or not, similar to a pass–fail situation.

2. P charts: This is the same as the nP chart, except the sample size varies.

3. C charts: C charts are used when there is more than one defect per unit and it is calculated as the number of defects produced when each unit has the same opportunities for defects (e.g., a constant sample size).

4. U charts: Similar to C charts, U charts are used to track defects per unit, but the unit is of varying size. For example, if the number of errors on an invoice is measured, and the fields of data on the invoice are constant from invoice to invoice, then a C chart is appropriate. A U chart is appropriate for measuring the number of errors in outgoing correspondence that is of variable size and contains data that differ among letters.

7.5.5.2 Constructing and Interpreting Control Charts

The steps to construct a control chart, along with an example for an X-bar and R chart, are described below. I recommend using statistical software, such as Minitab, to develop control charts. Even with charting software, the focus of effort in control charting is on the data collection, not actual generation of the charts. Automated data collection for control charting is common in manufacturing and rare in transactional services. Expect to spend the majority of the effort in developing a sound measurement and data collection plan.

1. Select the appropriate output measure according to the data collection plan. Customer CTQs should be tracked if possible. Depending on the output measure, select the appropriate control chart as described in Section 7.5.5.1.

2. Select the appropriate subgroup size. Subgroup sizes of four or five are most typically used. Approximately 100 data points, or 20 to 25 subgroups, need to be collected in order to establish the baseline center line and upper and lower control limits.

3. Calculate the center line and upper and lower control limits
4. On graph paper, draw the center line and the control limits. For X-bar charts, draw both the averages chart and the range or standard deviation chart. Draw two dotted lines to indicate 1 and 2 standard deviations from the center line. (Remember, the upper and lower control limits are set at 3 standard deviations from the center line, so divide the area between the control limits and the center line into three identical zones representing 1, 2, and 3 standard deviations from the mean center line.) Label the center line and control limits and provide a description of the data being tracked and the time period.
5. Plot the points.
6. Interpret the charts. Signs of an out-of-control situation are:
   a. One or more plots outside the upper or lower control limits.
   b. At least two successive points fall in the area beyond 2 standard deviations from the center line in either direction.
   c. At least four successive points fall outside the 1 standard deviation line.
   d. At least eight consecutive values fall on either side of the center line.
7. When an out-of-control situation occurs, investigate further to determine if an assignable cause is at work. Determine the root of the special cause and take corrective action. If no special cause can be identified, continue monitoring and take action as necessary.

Because of the probabilistic nature of control charts, it is possible that a Type I error has occurred. Type I errors are when an out-of-control signal is given but there is no special cause at work. According to the normal distribution, on which control charts are based, this will happen approximately 0.27% of the time. When a Type I error does occur, it is especially troublesome because time and effort is devoted to finding a special cause when there is none. On the other hand, a Type II error occurs when there is an out-of-control situation but the chart fails to detect it. When this happens, the process will be producing defects but you will be unaware of it.

7.5.5.3 Normality Assumptions

Much ado has been made regarding the requirement of normally distributed data for continuous data control charts. However, Wheeler and Chambers have demonstrated, through simulations using different types of continuous distributions,[16] that even if the data are non-normally distributed, control charts are still effective as a management tool.[17] This is because data points outside of 3 standard deviations from the mean for most distributions are still a rare event according to the laws of probability. Whereas only normally distributed data will exhibit the chance occurrence of 0.27% falling outside 3 standard deviations, other non-normal distributions will fall outside the 3 standard deviations limits at most 4% (chi-square and exponential), and many fall within 2% (uniform, right triangle, and Burr). This means that even significant departures from normality will have little effect on the function of control charts to signal uncontrolled variation. Shewhart developed control charts as a management tool to predict future behavior based on past experience. Their robust application in many real-world situations (where data is not always normally distributed) is the reason they have been successfully used in quality improvement for 75 years.

7.6 SIX SIGMA MEASUREMENTS

Almost all who are familiar with the Six Sigma approach can readily recite the specific measurement of it: 3.4 defects per million opportunities, abbreviated as DPMO (Table 7.8). But what does this really mean? When a team is involved in a Six Sigma project to reduce the cycle time of processing invoices, what does 3.4 DPMO have to do with it? Can they translate their cycle time data (e.g., average of 5.6 days with a standard deviation of 3.5 days) in terms of 3.4 DPMO? This section explores this question and others.

Because Six Sigma is a customer-focused approach, the first step in calculating sigma levels is to identify the customer CTQs, which were defined earlier as process or service attributes that must be met to satisfy the customers' requirements. Once these are identified through VOC, we can identify the defects (D) as any situation in which the customers' CTQs are not met. The next step is to identify the unit of service (U). This could be a product, such as a billing statement, or a transaction, such as a phone call. Then we identify the number of opportunities for defects (OP) for each unit of service and the total number of opportunities, which is the number of units multiplied by the opportunities per unit (TOP). The defects per unit (DPU), defects per total opportunities (DPO), and DPMO are all measurements of defects per unit. From the DPMO, the yield (Y) and process sigma levels are calculated according to the sigma conversion table (Table 7.9). The yield is the percentage of units produced without defects. Let us look at an example of a process to bill a customer for an outpatient procedure:

1. The customer CTQ has been identified as error-free billing. There are 120 different fields on the billing statement, each of which could be a defect. Therefore, the OP = 120
2. A defect is defined as any field that is missing information or has incorrect information.
3. Sampling has revealed that there are 2.0 defects for every 10 bills processed.
4. Total opportunities for defects (TOP) in ten bills = $120 \times 10 = 1200$.
5. DPU = D $\div$ U = 2 $\div$ 10 = 0.2.
6. DPO = D $\div$ TOP = 2 $\div$ 1200 = 0.001667 DPMO = DPO $\times$ 1,000,000 = 0.001667 $\times$ 1,000,000 = 1666.667.
7. Refer to the sigma conversion table (Table 7.9). Look for 1,667 in the DPMO column.
8. Identify the sigma level and long-term yield. Because 1,667 does not appear in the DPMO column, interpolate to arrive at a sigma level of 4.45 with a yield of approximately 99.8%.
9. To arrive at rolled throughput yield (RTY), or final yield, multiply the yields from each step of the process. For example, suppose the process of sending billing statements involves three steps: (1) producing accurate

TABLE 7.8
Six Sigma Measurement Terms[18]

Term	Symbol	Description
Units	U	Base unit of process or product that can be quantified
Defects	D	Any part of process or product that is unacceptable to the customer; does not meet specifications
Opportunity	OP	Total number of chances per unit to have a defect; must be measurable
Total opportunities	TOP	Number of units multiplied by number of opportunities; U × O
Defects per unit	DPU	Number of defects divided by number of units; D ÷ U
Defects per total opportunities	DPO	Number of defects divided by total number of opportunities; DPU ÷ O = D ÷ (U × O) or D ÷ TOP
Defects per million opportunities	DPMO	Number of defects divided by total number of opportunities, then multiplied by 1,000,000; DPO × 1,000,000
Yield	Y	Percentage of units produced with no defects, calculated as total number of opportunities minus total number of defects, divided by total number of opportunities, then multiplied by 100; also called first-pass yield; [(TOP − D) ÷ TOP] × 100
Rolled throughput yield	RTY	Number of process steps multiplied by yield of each process step; also called final yield

statements, (2) printing the statement within 24 h of the monthly cutoff, and (3) mailing it to the customer within another 24 h. We calculate the yield at each step in the process. Step 1, producing accurate statements, has a yield of 99.8%, and step 2, printing the statement within 24 h, has a yield of 97.3%. In step 3, 95.6% of statements are mailed to the customer within an additional 24 h. The RTY is calculated as .998 × .973 × .956 = .928, or a 92.8% yield. According to the sigma conversion measure in Table 7.9, this equates to a sigma level of 2.99, which is the final process yield of the billing statement process.

Note that the sigma conversion table provides long-term process yield. The "correction factor" that helps distinguish between long and short term is what is known as the 1.5 sigma shift. The theory behind the 1.5 sigma shift is that over the long run, process means tend to shift by 1.5 sigma. For example, in measuring the talk time of a phone call, the mean value will probably drift from a short-term average of 180 sec (standard deviation of 20 sec) to a long-term average of 150 to 210 sec (180 sec ± 1.5 × 20 sec). However, it is typical to calculate sigma levels on short-term data. To measure long-term process capability in terms of process sigma measurements, the 1.5 sigma shift is added to the short-term process capability or sigma level. In statistical terms, a six sigma process is capable of producing only 2 defects per billion opportunities (DPBO) rather than the 3.4 DPMO we typically define as six sigma capability. In other words, the 3.4 DPMO is the proportion defective outside of the normal curve specification limits, assuming that the process mean has shifted by 1.5 sigma, which is likely to happen in the long run. With no

TABLE 7.9
Sigma Conversion Table

Process Sigma Level	Long-Term Yield	DPMO
1	30.2%	697,600
1.1	34.0%	660,100
1.2	37.9%	621,400
1.3	41.9%	581,800
1.4	45.8%	541,700
1.5	49.9%	501,300
1.6	53.9%	461,100
1.7	57.9%	421,400
1.8	61.7%	382,600
1.9	65.5%	344,900
2	69.12%	308,800
2.1	72.56%	274,400
2.2	75.79%	242,100
2.3	78.81%	211,900
2.4	81.59%	184,100
2.5	84.13%	158,700
2.6	86.43%	135,700
2.7	88.49%	115,100
2.8	90.32%	96,800
2.9	91.92%	80,800
3	93.31%	66,800
3.1	94.52%	54,800
3.2	95.54%	44,600
3.3	96.41%	35,900
3.4	97.13%	28,700
3.5	97.72%	22,800
3.6	98.21%	17,900
3.7	98.61%	13,900
3.8	98.93%	10,700
3.9	99.18%	8,200
4	99.379%	6,210
4.1	99.533%	4,660
4.2	99.653%	3,470
4.3	99.744%	2,550
4.4	99.813%	1,870
4.5	99.865%	1,350
4.6	99.903%	965
4.7	99.9312%	685
4.8	99.9516%	485
4.9	99.9663%	340
5	99.97673%	233
5.1	99.98408%	159

TABLE 7.9 (continued)
Sigma Conversion Table

Process Sigma Level	Long-Term Yield	DPMO
5.2	99.98922%	108
5.3	99.99273%	72.4
5.4	99.99519%	48.1
5.5	99.99683%	31.7
5.6	99.99793%	20.7
5.7	99.99867%	13.35
5.8	99.99915%	8.55
5.9	99.99946%	5.42
6	99.99966%	3.4

---

assumption of process shifts, 3.4 DPMO represents a 4.5 sigma level rather than a 6 sigma level. Table 7.9 provides a quick and easy reference table to convert yield (accuracy) rates to DPMO and to process sigma levels.

Regardless of the assumption of process shifts of 1.5 sigma, what is important is not so much the precise measurement of 3.4 DPMO or 2.0 DPBO, but the goal of continuous improvement. In manufacturing, where precise measurements of diameter dimensions and tolerances are made with devices called micrometers and bore gages, it is far more important to understand the long-term vs. the short-term parts per million outside of specifications. However, in service environments, it may not always be worth the time and effort to get to this level of detailed measurement. What is important is understanding the customers' CTQs and measuring the ability of your processes to meet those CTQs — because of the human element involved in the critical moments of truth, whether the phone call is answered in 30.8885 sec or 30.9005 sec is not material. Each situation is different, and Black Belts, Green Belts, and Six Sigma teams are constantly challenged to understand the level of detail necessary to continuously and profitably meet and exceed customer expectations.

7.7 COST OF POOR QUALITY

One of the differences between a Six Sigma quality improvement approach and many prior quality programs is the emphasis on financial returns. Because Six Sigma is a business initiative rather than a quality initiative it makes sense that financial returns of Six Sigma projects are measured. And, as Joseph Juran pointed out many years ago, the language of business is money.

The cost of quality (COQ) refers to the costs associated with preventing defects. The cost of poor quality (COPQ) refers to the costs associated with failure in the process. We will examine preventive and detection costs in terms of COQ and failure costs as COPQ.

7.7.1 Preventive Cost of Quality

Preventive costs are the costs of doing it right the first time. They are the least costly of the three categories because costly mistakes are prevented from happening in the first place. Examples include:

- Design for Six Sigma
- Process improvement
- Hiring
- Education
- Training
- Quality reward and recognition programs
- Planning
- Supplier certification
- Customer surveys
- Statistical process control
- Preventive maintenance

Preventive costs prevent failures, thus ensuring that customer requirements are continually met. Preventive costs are also referred to as the cost of conformance in manufacturing and the cost of compliance in transactional services.

7.7.2 Appraisal Cost of Quality

Appraisal costs, also called detection costs, are the costs to find things that were not done right the first time. Examples include:

- Inspections
- Authorizations
- Audits
- Proofreading
- Customer surveys
- Statistical process control
- Spell checking
- Call monitoring

Appraisal costs, like preventive costs, are grouped under the cost of compliance; they are the costs of activities needed to evaluate compliance to customer requirements. Appraisal costs, although higher than preventive costs, are less than the cost of failure, because appraisal activities prevent the process from producing defective products or services.

7.7.3 Failure Cost of Poor Quality

Also known as the costs of noncompliance, failure costs are the costs of not doing things right the first time. There are two categories of failure costs: internal failure and external failure. Both are very costly.

Internal failure costs include the cost of fixing mistakes, better known as rework. These are failures that are found and fixed before getting to the customer and the costs associated with that. Internal failure costs are usually associated with the "re-" words, such as rectification, rework, reinspection, review, and rewrite. Examples include:

- Scrap (wasted paper, machines, toll charges)
- Excess labor
- Overtime
- Staff absenteeism, turnover, and health costs

External failures are the worst of all the quality costs. These are the failures that are detected by customers and the costs of activities associated with meeting customer requirements after delivery of the product or service to the customer. Examples include:

- All of the costs of internal failures
- Warranty costs
- Labor costs to investigate
- Regulatory investigative costs
- Lawsuits
- Lost business
- Product recalls
- Service failure recoveries
- Credit losses

7.7.4 Cost of Quality Profiles

In the typical service company, COQ accounts for 30 to 50% of revenues. It is slightly lower in manufacturing companies — 15 to 35% of revenues. As shown in Figure 7.21, the "good" kind of quality cost (preventive) typically accounts for only 10% of the total quality costs, whereas failure accounts for a whopping 65%.[19] These figures are before any type of Six Sigma quality costing system has been implemented. Through application of Six Sigma projects aimed at building preventive quality into the processes, the results illustrated in Figure 7.21 can be achieved after 2 to 3 years.

The COQ has a significant impact on profits. Consider, for example, a company with $100 million in revenues and an operating income of $10 million. Assume that 30%, or $30 million, of revenues are quality costs. If this company could cut its COQ by only 20%, it would increase its operating income by $6 million, or 60% of operating income. This example clearly demonstrates the importance of identifying and tracking quality costs and increasing the percentage of preventive quality costs vs. failure quality costs. By decreasing the occurrence of defects and the need for rework, Six Sigma projects can be successful in decreasing the costs of poor quality and the resulting percentage of revenue, thus increasing bottom line profitability.

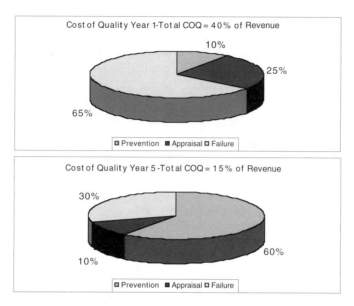

FIGURE 7.21 Cost of quality (COQ) in year 1 (before Six Sigma) and 5 years into Six Sigma.

7.8 PROBABILITY DISTRIBUTIONS

We touched on probability distributions in Section 7.2 with discussion of the normal distribution, which is by far the most common distribution in nature and in business applications. This section examines the numerous other data distributions that are often used in Six Sigma projects. Recall from earlier sections that probability and statistical inference are tools used for predicting what the distribution of data should look like in a particular situation, such as the distribution of heights of randomly selected individuals. Random phenomena are not haphazard: they display an order that emerges only in the long run and is described by a distribution. Figure 7.22 summarizes the more common distributions' probability functions and their applications.

7.8.1 EXPONENTIAL DISTRIBUTION

The exponential distribution provides the distribution of time between independent events occurring at a constant rate. This distribution is used when we are interested in the distribution of the interval (measured in time units such as minutes) between successive occurrences of an event. A common application in Six Sigma projects is queuing studies, especially to forecast incoming call volume in call centers. The continuous exponential probability distribution has an interesting relationship to the discrete Poisson distribution (see Section 7.8.4) for this situation and any situation involving the occurrence of events ordered in time. The Poisson distribution describes the arrival rate of the incoming calls in a particular time unit and the exponential distribution describes the time between call arrivals. The exponential

Distribution	Shape	PDF
Exponential μ = mean μ^2 = variance Provides the distribution of time between independent events occurring at a constant rate – used in mean time to failure and queuing applications	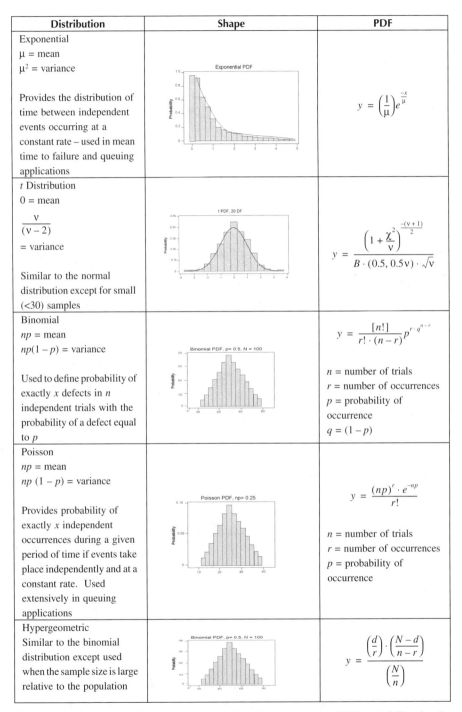	$y = \left(\dfrac{1}{\mu}\right)e^{\frac{-x}{\mu}}$
t Distribution 0 = mean $\dfrac{v}{(v-2)}$ = variance Similar to the normal distribution except for small (<30) samples		$y = \dfrac{\left(1 + \dfrac{\chi^2}{v}\right)^{\frac{-(v+1)}{2}}}{B \cdot (0.5, 0.5v) \cdot \sqrt{v}}$
Binomial np = mean $np(1-p)$ = variance Used to define probability of exactly x defects in n independent trials with the probability of a defect equal to p		$y = \dfrac{[n!]}{r! \cdot (n-r)} p^r \cdot q^{n-r}$ n = number of trials r = number of occurrences p = probability of occurrence $q = (1-p)$
Poisson np = mean $np\,(1-p)$ = variance Provides probability of exactly x independent occurrences during a given period of time if events take place independently and at a constant rate. Used extensively in queuing applications		$y = \dfrac{(np)^r \cdot e^{-np}}{r!}$ n = number of trials r = number of occurrences p = probability of occurrence
Hypergeometric Similar to the binomial distribution except used when the sample size is large relative to the population		$y = \dfrac{\left(\dfrac{d}{r}\right) \cdot \left(\dfrac{N-d}{n-r}\right)}{\left(\dfrac{N}{n}\right)}$

FIGURE 7.22 Common distributions used in Six Sigma projects. (PDF, probability density function)

distribution is also useful in reliability studies to determine the mean time between failures.

7.8.2 *t* DISTRIBUTION

The *t* distributions were discovered in 1908 by William Gosset, a chemist and statistician employed by the Guinness brewing company. Guiness required its employees to publish research anonymously, and Gosset chose to sign his papers "student." Thus the *t* distribution is often referred to as the student's *t* distribution.

t Distributions are a class of distributions similar to the normal distribution — symmetric and bell shaped. However, the spread is greater than for the standard normal distribution. When we speak of a specific *t* distribution, we must specify the degrees of freedom (v). The degrees of freedom are determined from the sample size: for a small sample size, the *t* distribution is flatter, with more data in the tails, and as the sample size increases beyond 50 observations, the *t* distribution is similar to the standard normal distribution. The *t* distribution is used like the normal distribution (to test and estimate population means), but in situations where the sample size is small. It is also used to determine if two population means are different — a very useful test in Six Sigma projects. The use of the *t* test to determine differences between two populations is discussed further in Chapter 8.

7.8.3 BINOMIAL DISTRIBUTION

The binomial distribution is used when there are only two outcomes of interest, such as heads or tails, pass or fail, and yes or no. It provides the probability of exactly x defects in n independent trials, with the probability of a defect equal to p. For example, the binomial distribution can be used to determine how many defective units are expected in a sample of 10 units if the defect rate is 1 out of 50.

7.8.4 POISSON DISTRIBUTION

The Poisson distribution is a discrete distribution similar to the binomial. It gives the probability of exactly x independent occurrences during a given period of time if events take place independently and at a constant rate. One of the most useful applications of the Poisson distribution is in the field of queuing theory. In many situations where queues occur it has been shown that the number of people joining the queue in a given time period follows the Poisson model. Examples include arrival rates for incoming calls, emergency room visits, and bank teller lines. As discussed in Section 7.8.1, the exponential distribution is used to describe the time spacing between arrivals.

7.8.5 HYPERGEOMETRIC DISTRIBUTION

The hypergeometric distribution is also similar to the binomial distribution, except it is used when the sample size is large relative to the population. Generally, when the sample size is greater than 10% of the population, the hypergeometric distribution should be used instead of the binomial.

7.9 TECHNICAL ZONE: MEASUREMENT SYSTEM ANALYSIS

When measuring process quality, it is important to understand the concept of *measurement system analysis*. Process variation typically stems from two distinct sources — *actual process variation* and *measurement variation* (Figure 7.23). This is an important concept during the Measure Phase because the project team could identify that a process has unacceptable variation and go about identifying root causes in the Analysis Phase and develop solutions in the Improve Phase and then in the Control Phase, fail to see expected improvements. How could this happen? The source of variation could be in the way the process variation is measured — the actual variation is in the measurement system, not in the process. Measurement system analysis is a method for identifying and separating out the variation due to the measurement system from actual process variation. Only through this process are true root causes of process variation identified and effective solutions developed.

7.9.1 GAUGE R & R

As the unit of work decreases in size, the need for a more precise measurement system increases. That is, the smaller the measurement unit of scale, the more precise the measurement system needs to be. This is true whether it is a physical part or a unit of time. If we are measuring the diameter of a paper cup to the nearest inch, everyone who measures the cup will arrive at the same measurement — it is fairly obvious whether it is 1 in., 2 in., or larger. If we were to measure the same cup diameter to the nearest half-inch, we would probably still end up with the same measurement regardless of who did the measuring. However, if we needed to measure the diameter of the cup to the nearest $1/8$ or $1/16$ in., we would start to see some variation in the measurement results, depending on who took the measurements or, if performed by one person, when the measurements were taken. This concept is known as gauge R & R, which stands for gauge repeatability and reproducibility. Gauge refers to the instrument or device used for making measurements.

Repeatability is the variation observed when the same person measures the same thing with the same device; in other words, it is the variation in the measurement device. Reproducibility is the variation observed when different people measure the same thing with the same instrument and is therefore the variation in the people

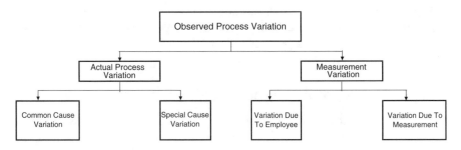

FIGURE 7.23 Sources of observed process variation.

taking the measurements. Together, these two make up the gauge R & R, which is the variation due the measurement system rather than the actual process.

A subset of repeatability is calibration. Calibration is the process of making sure the measurement device, the gauge, is accurate. In other words, does the instrument measure what it is supposed to measure? This is discussed further in the next section.

7.9.2 SERVICE VS. MANUFACTURING

In Section 7.5, we discussed the difference between achieving a six sigma quality level in a manufacturing process vs. a service process. Generally, in the manufacture of component parts, whether it is a jet engine turbine part or a flashlight on/off switch, consistency (lack of variation) is very important. When component parts come together in subassembly, lack of tight tolerances results in product defects. The same concepts apply in transactional services, but usually on a different scale. Because inputs into service processes are in large part actions taken by employees, the required precision of the measurement scale is not as well defined. In measuring the cycle time of a new product development process that is on average 12 to 18 months, do we really need to measure to the minute? Of course not. However, in measuring the average time it takes for an agent to answer an inbound phone call, every second seems to make a difference (at least to the customer who is waiting). The point is there are many more applications of measurement system analysis in a manufacturing environment, where tolerances to the .001 in. can be critical to producing a defect-free product. Measuring down to the minute or even to the second in a service process may not be worth the time and effort except in special cases, such as the time to answer a call or perhaps the time to respond to an emergency call.

Nonetheless, numerous applications of the calibration of measurements in a service environment exist. A common example is the process of monitoring phone calls for the purpose of measuring the quality of the call center agent by quality assurance analysts. It is typical, and highly recommended, that calibration sessions are routinely held to ensure the measurement system is consistent among different quality assurance analysts. A common method to calibrate the measurement of quality in a call is to have several quality assurance analysts rate the same call according to the company's predetermined quality standards. The results are compared and discussed among the group and a consensus achieved on how to consistently rate the quality of the agent. Another common application of calibration deals with performance reviews. Do all managers agree on what constitutes "exceeds expectations" vs. "meets expectation"? Calibration sessions similar to those for call monitoring can be held to ensure consistent application of the performance standards among all managers.

7.10 TECHNICAL ZONE: PROCESS CAPABILITY

The ability of a process to meet customer specifications is called *process capability* and is measured through the process capability index, Cp. Cp is a measure of the allowed variation or tolerance spread compared to actual data spread. The actual process output is compared to the customer specifications and judgments are made as to whether the process is capable of meeting those specifications.

7.10.1 CALCULATING PROCESS CAPABILITY

The formula for Cp is:

$$Cp = (USL - LSL)/6\sigma \tag{7.8}$$

where USL = upper specification limit and LSL = lower specification limit.

Cp measures customer specifications in terms of acceptable spread or dispersion distances, with no reference to the target or center value of the data. For example, customer specifications for processing a bank wire may state that the wire must be sent within 4 to 6 h from the time the wire request is made. Hence, 6 h is the upper specification limit and 4 h is the lower specification limit. The tolerance level is the difference between the upper and lower specification limits, which in this case is 2 h. Figure 7.24 provides a graphical illustration of this example.

$$Cp = (6 - 4)/(6 \times 0.32) = 2/1.92 = 1.04 \tag{7.9}$$

What does this mean? It means that the actual performance of processing bank wires will meet the customer requirements most of the time. If the customer specifications exactly equal the natural process limits, this measure is 1.0.

When we measure process capability, we are comparing actual process performance (voice of the process) to the customer's requirements (voice of the customer). The actual process performance is calculated as the mean value ±3 standard deviations from the mean, the same as the control chart performance measurement. The upper and lower control limits are called the *natural process limits*. Recall that for normally distributed data, 99.73% of all observations will fall within the natural process limits. The distance between the upper control limit and the lower control limit is six sigma, because the distance is three sigma on either side of the mean. As long as the specifications are within the natural process limits, the process will meet customer specifications; otherwise, some process output will be outside of the specifications. This is illustrated in Figure 7.24.

What if, in addition to an upper and lower specification limit, we wanted a target value for processing bank wires? The measurement for this situation is called Cpk and is like Cp, except that it considers both the mean and the spread of the data. If, for example, the customer specifications required bank wires to be processed on average within 5.0 h, with a lower limit of 4.0 h and an upper limit of 5.5 h, the formula is:

$$Cpk = min\ [USL - \mu/3\sigma, \mu - LSL/3\sigma]$$

$$= min\ [0.52, 1.04]$$

The Cpk in this example is 0.52, because it is the minimum of the two measures. This example is graphically represented in Figure 7.25.

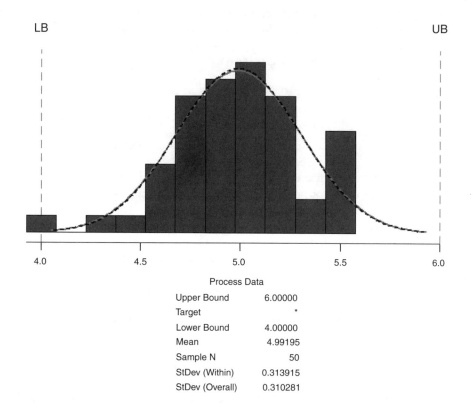

FIGURE 7.24 Process capability (Cp) of bank wire processing time. (LB, lower bound; UB, upper bound)

7.10.2 THREE SIGMA CONTROL LIMITS VS. SIX SIGMA SPECIFICATION LIMITS

A distinction needs to be made between control limits and specification limits. As we saw in Section 7.5 on control charts, control limits are typically set at ±3 standard deviations (sigma units) from the mean. We know that ±3 standard deviations from the mean measures the natural process limits, or the voice of the process. When we discuss a six sigma level, we mean a customer specification measurement of ±6 standard deviations (sigma units) from the mean. Specification limits indicate acceptable vs. defective units. Control limits are statistical in nature and a function of how disperse the actual process is. They are called control limits because you use them to *control* the process (i.e., when to take action and when not to). Note the difference: control limits are the *voice of the process* and six sigma limits are the *voice of the customer*. Achieving a six sigma level of performance means that the process is producing 3.4 DPMO according to customer specifications (Figure 7.26).

Thus, for a process to reach a six sigma level of performance, one must reduce the natural variation of the process from common cause. Specifically, to take a process that is in a state of statistical control (i.e., producing only 0.27%, or.27 parts

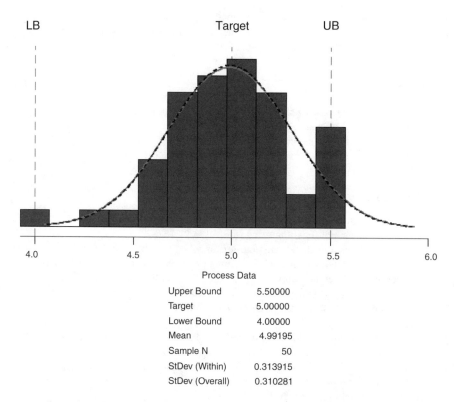

Process Data

Upper Bound	5.50000
Target	5.00000
Lower Bound	4.00000
Mean	4.99195
Sample N	50
StDev (Within)	0.313915
StDev (Overall)	0.310281

FIGURE 7.25 Process capability (Cpk) of bank wire processing time. (LB, lower bound; UB, upper bound)

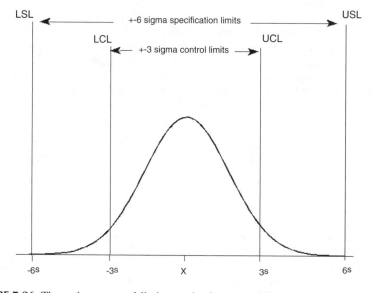

FIGURE 7.26 Three sigma control limits vs. six sigma capability.

per 100 opportunities) to a six sigma level of performance (i.e., producing only 0.0000034 parts per 100 opportunities), the variation must be improved by a huge factor (almost 19,650)!

REFERENCES

1. Cusack, M., *Online Customer Care: Strategies for Call Center Excellence*, ASQ Quality Press, Milwaukee, WI, 1998, p. 49.
2. The height of a normal curve is defined as:

$$\frac{1}{\sqrt{2\pi\sigma^2}} \cdot e^{-(x-\mu)^2/(2\sigma^2)}$$

   where μ is the mean, σ is the standard deviation, π is the constant 3.14159, and e is the base of natural logarithms and is equal to 2.718282.
3. For the purposes of this discussion, it is assumed that the underlying distribution of the data is normal, which is a fair assumption given the central limit theorem, which states that for even non-normal populations, the sampling distribution from the target population is approximately normally distributed.
4. Zikmund, W., *Business Research Methods*, 4th ed., Dryden Press, New York, 1994, p. 409.
5. Zikmund, W., *Business Research Methods*, 4th ed., Dryden Press, New York, 1994, p. 282.
6. Camp, R.C., *Benchmarking: The Search for Industry Best Practices that Lead to Superior Performance*, ASQ Quality Press, Milwaukee, WI, 1989.
7. http://www.minitab.com, 2001.
8. http://www.SAS.com. 2001.
9. http://www.statgraphics.com, 2001.
10. http://www.stanford.edu/class/msande269/six_scholars_comparisons.html.
11. http://www.moresteam.com, 2001.
12. Breyfolge, F., III, *Implementing Six Sigma: Smarter Solutions Using Statistical Methods*, John Wiley & Sons, New York, 1999, p. 19.
13. Dr. Shewhart gave the following definition of statistical control: "A phenomenon will be said to be controlled when, through the use of past experience, we can predict, at least within limits, how the phenomenon may be expected to behave in the future." From Wheeler, D.J., and Chambers, D.S., *Understanding Statistical Process Control*, 2nd Ed., SPC Press, Knoxville, TN, 1989, p. 6.
14. http://asq.org/join/about/history/shewhart.html, 2001.
15. Shewhart, W., *Economic Control of Quality of Manufactured Product*, Van Nostrand Reinhold, New York, 1931.
16. The simulated distributions were uniform, right triangle, Burr, chi-square, and exponential. From Burr, I.W., *Elementary Statistical Quality Control*, Marcel Dekker, New York, 1979.
17. Wheeler, D.J. and Chambers, D.S., *Understanding Statistical Process Control*, 2nd ed., SPC Press, Knoxville, TN, 1989, p. 56.
18. Breyfolge, F., III, *Implementing Six Sigma: Smarter Solutions Using Statistical Methods*, John Wiley & Sons, New York, 1999, p. 140.
19. Hou, T-f., "Cost-of-Quality Techniques for Business Processes," ASQ FAQ Series on Cost of Quality, 1998, from a presentation at the 46th Annual Quality Congress, 1992.

8 Analyze Phase

8.1 OVERVIEW

During the Measure Phase, the Six Sigma team collects significant amounts of data in order to measure the current performance of the process and identify the process critical X's and Y's. The objective of the Analyze Phase is to confirm the hypotheses, or educated guesses, made during the Measure Phase as to the root causes of the problems. Only after the team has moved beyond symptoms of the problem to the true root cause can they develop effective solutions in the Improve Phase.

The Analyze Phase involves statistical analysis of collected data and, thus, use of statistical software, such as Minitab, is recommended. Excel offers statistical functions and tests through its Data Analysis ToolPak, which is an add-in found under the tools menu; however, specialized software is needed to take full advantage of the Analyze Phase tools.

A statistical approach to decision making involves moving from *data* to *information* to *knowledge*. At the beginning of a Six Sigma project, the team is busy collecting all sorts of data; however, the data are just a lot of numbers at this point. Through the measure and analyze tools, data become information with which hypotheses can be made. Information becomes knowledge when it is successfully used to make decisions and the lessons learned from the process are completed. Statistical tools move us from data to knowledge by helping us make decisions under uncertainty. Decision making must be based on objective facts and knowledge and statistics is the science that brings clarity to what is otherwise meaningless data (Figure 8.1).

The Analyze Phase covers the following topics:

1. Process analysis based on the process maps developed in the Measure Phase
2. Hypothesis testing, including a discussion of p values and alpha and beta
3. Statistical tests and tables
4. Statistical tools to measure relationships between variables

The team should exit the Analyze Phase with a clear understanding of the root causes of the problems that prevent the process from meeting the customer CTQs identified in the Define and Measure Phases.

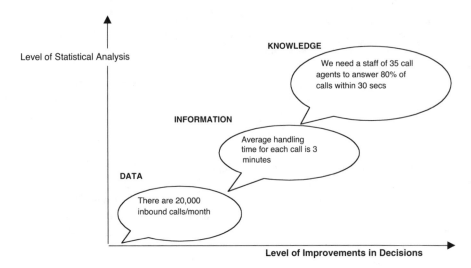

FIGURE 8.1 Statistical approach to decision making.

8.2 PROCESS ANALYSIS

Analyzing the process maps developed in the Measure Phase is one of the first steps in the Analyze Phase. This involves identifying:

1. Value added vs. non-value-added activities
2. Direct customer contact steps
3. Sources of process delays
4. The impact of policies on process performance

8.2.1 VALUE-ADDED VS. NON-VALUE-ADDED ACTIVITIES

The goal in *value-added analysis* is to identify the value-added steps vs. the non-value-added steps and eliminate as many non-value-added steps as feasible. An activity is adding value if it meets the following three criteria:

1. The customer is willing to pay for it
2. Inputs are transformed into outputs
3. It is done right the first time (no rework)

Process steps that meet the above criteria are essential to the process. In addition, steps that help the value-added steps work better, known as "value-enabling steps," are important. The team needs to be primarily concerned with the other steps, or the non-value-added steps.

8.2.2 CUSTOMER CONTACT POINTS

In a 1992 *Harvard Business Review* article on process analysis, titled "Staple Yourself to an Order,"[1] the authors describe the journey of a customer order moving through the order and fulfillment process. Through this imaginary process, the authors open our eyes to the experience of our external customers as they move through our internal processes. Clearly highlighted are the "moments of truth" — those direct experiences between the company and the customers that shape how the customers perceive the business and their desire to continue with the company.

In performing a value-added process analysis, it often helps to imagine yourself as an order (or a claim, request, dispute, etc.) moving through the process. What steps are you, as the customer (order), willing to pay for? Certainly not something that is done incorrectly the first time and therefore has to be reworked. How about moving from one person to the next for another round of approvals, after waiting in an in box for more than 2 weeks? This imaginary journal is not only fun (or maybe painful), it also helps the team to clearly identify the value-added steps.

8.2.3 PROCESS DELAYS

After the value-added analysis is completed and the customer contact points identified, the total work time of the process compared to the total process cycle time needs to be determined. Work time is defined as the amount of time that something is actually being done to the process step, whether value added or non-value added. The cycle time is measured from the beginning of the process, usually the customer request, to the completion point, which is usually when the customer request is fulfilled. The difference between work time and cycle time is the *delay time*.

Six Sigma teams are often astonished at how little work time is spent on the process compared to how long it takes to complete the process cycle. Typically, only 10 to 20% of the total cycle time is work time — the other 80 to 90% is idle time waiting for something to happen. Sources of delays and, thus, potential areas for improvement may include the following:

1. Approvals, especially multiple levels of approvals.
2. Any hand-off situation, because hand-offs create queues and queues mean delays.
3. Bottlenecks, because bottlenecks create queues.
4. Defects, because defects create rework, which adds time to the process, thus delaying completion.
5. Inspections or appraisal activities. Not only do inspections add unnecessary time to the process, they do not guarantee accuracy. Studies have shown that only approximately 85% of defects are identified through inspection. You cannot inspect quality into a process or a product.
6. Unclear operational definitions. As discussed in Chapter 7, clear, unambiguous instructions through operational definitions are needed to ensure that everyone involved in the process has the same interpretation of the work to be accomplished.
7. Unclear accountability for the work. If everyone is responsible, no one is responsible.

8.2.4 EFFECT OF POLICIES ON PROCESS PERFORMANCE

Significant process delays are often a mixed blessing to a Six Sigma improvement team. Although the delays may be the lowest of the "low-hanging fruit" in achieving dramatic improvements in cycle time, the single largest source of delays are typically approvals — specifically, waiting for approvals. Approvals are usually in the domain of policy rather than process, which leads us to a discussion on the role of company policies and how they affect process performance.

On numerous occasions, I have witnessed the frustration of Six Sigma teams who have correctly identified the root cause of the problem, and to quote Oliver Perry, "We have met the enemy and it is us!" Although changing the process is difficult, it may be a cakewalk compared to challenging company policy. However, the most dramatic improvements are often achieved through changing outdated or non-customer-friendly policies. A good example is a retailer's return policies. Because the importance of customer satisfaction in keeping a business afloat has become well known over the past two decades, most retailers have accurately identified their return policy as a customer CTQ. Most customers do not even ask about return policies; the assumption is that if for some reason the product does not meet their satisfaction after the purchase, they can return it for a refund or credit. However, some retailers have not moved into the new era of seeing customer satisfaction as a competitive advantage; they make it difficult to return an item, regardless of the cause. If a Six Sigma team correctly identifies a liberal return policy as a customer CTQ and, furthermore, identifies the current return policy as dissatisfactory and time consuming, but they cannot affect the company's return policy, everyone loses: the customer, the company, and the Six Sigma team. The fact that Six Sigma is a business initiative that is owned by the top executives of a company provides encouragement that policy, as well as process, is open for debate in ultimately achieving the goal of bottom line improvements. This said, company policy is an area that the project team should be willing to review and, if necessary, challenge.

8.2.5 THE THEORY OF CONSTRAINTS

The theory of constraints is a process analysis approach developed by Dr. Eli Goldratt.[2] The essence of the theory is that a process is composed of many process steps, each with their own capacity for transforming inputs into outputs. The process can produce only as fast as its slowest process step. Therefore, to increase the output of the total process, identify the slowest step (also called the "bottleneck" or the "constraint") and fix it. Fixing means providing greater capacity to the slowest process step (by increasing the productivity of the step or finding alternative ways to perform the step) so that the total output of the system (process) is increased.[3] Goldratt refers to fixing the slowest process step as "elevating" the constraint or lifting the workload of the constraint through alternative methods. The team identifies and elevates each constraining step in succession.

Goldratt popularized his theory of constraints in a very successful book called *The Goal*. The goal of any business is profit, and there are only three ways to achieve increased profits:[4]

1. Increase throughput (the total output of the process)
2. Decrease expenses
3. Decrease inventories

The message of the book is relayed through the story of a factory manager who is struggling with increasing inventories, cycle times, and back orders, and decreasing profits. Listening to his eccentric but wise former college professor, the factory manager is able to turn the factory around using the theory of constraints.

8.3 HYPOTHESIS TESTING

Recall from Chapter 7 that inferential statistics is used primarily for two purposes: point estimation and hypothesis testing. We covered point estimates by calculating the mean, median, standard deviation, variance, and other statistics, along with confidence intervals and confidence levels. This section deals with hypothesis testing, which is testing assumptions about the data to assist in moving from data to information to knowledge; in other words, from uncertainty to some level of certainty (see Figure 8.1). Sample statistics, unlike population parameters, are based on probabilities, and thus, we can never be absolutely certain in our conclusions; we can say that there is a very high probability that our conclusions are correct.

Hypothesis testing is stated in the form of a null hypothesis and an alternative hypothesis. The null hypothesis asserts that the data belong to one population and the alternative states the data belong to a different population. The following equation illustrates a hypothesis test for whether a sample mean is equal to the population mean:

$$H_0 = \mu$$

$$H_1 \neq \mu \tag{8.1}$$

Many different statistical tests apply in hypothesis testing. The ones used most often in Six Sigma are covered in this chapter. They are the Z test, the t test, the F test, and the chi-square test. The type of test used in hypothesis testing depends on whether the data input (critical X's) and output (critical Y's) are continuous or discrete, and the number of variables involved in the testing, as follows:

- *Univariate* statistical tests, which test one variable against a target value, answer the question, "Is the sample mean different than the population mean?"
- *Bivariate* statistical tests, which test for differences between two variables, answer the question, "Are these two sample means significantly different from each other?" That is the same as asking, "Is there a difference in population means for the two samples?"
- *Multivariate* statistical tests are similar to bivariate statistics but test for differences between more than two samples.

We conduct statistical tests because we are dealing with a sample of data drawn from a larger population. Hypothesis testing involves making conclusions on whether a sample set of data is representative of, or belongs to, the population of interest.

For univariate data, the null hypothesis (H_0) asserts that the sample data are representative of the population from which they were drawn (same mean and standard deviation), and the alternative hypothesis (H_1) states that the sample data are not representative of the greater population (different mean and standard deviation). For bivariate data, the null hypothesis (H_0) asserts that the two sample data sets are from the same population source (same means and standard deviations), and the alternative hypothesis (H_1) states that the two samples come from two different population sources (different means and standard deviations). Multivariate hypothesis testing is similar to bivariate, except it tests that more than two sample data sets either originate from the same population (H_0) or are from two or more different populations (H_1).

Note that in all of these hypothesis tests, the null hypothesis states there is no difference between the samples, whereas the alternative states there is a difference. We begin by assuming the null hypothesis is true and set out to prove that is it wrong. The next section provides insight on how to draw conclusions from the hypothesis testing to determine if there actually is a difference.

8.3.1 STATISTICAL SIGNIFICANCE

A common statement regarding statistical tests is, "Based on the results of the hypothesis testing, there is a statistically significant difference between the two sample data sets." What exactly does this mean? The answer is provided in this section.

The process of hypothesis testing is calculating an appropriate test value (e.g., Z statistic) based on actual data and then comparing the test statistic to a critical value in the assumed, or theoretical, distribution of the data (e.g., normal distribution). From earlier discussions on the measurement error (E), we know it is highly unlikely that a sample statistic, such as the mean, will exactly equal the population mean for a given distribution of data. However, if the difference between the calculated statistic and the theoretical statistic found in statistical tables is great enough not to occur by chance alone, we can state with some level of confidence that we reject the null hypothesis of no difference and conclude there is a statistically significant difference between the two samples. We started out by asserting there was no difference (H_0) and proved it wrong.

The level of confidence we have in the results of the statistical test is based on a predetermined value called alpha, also called the significance level. Common values of alpha are 1%, 5%, and 10%. With an alpha of 5%, we assert that we will reject the null hypothesis if the obtained statistic is one that would occur only 5 times out of 100 (5%), which is considered a rare chance occurrence. Note that this means we will correctly fail to reject the null hypothesis 95 times out of 100, but will mistakenly reject the null hypothesis 5 times out of 100. The probability of mistakenly rejecting the null hypothesis is exactly where we set the alpha value. We use the term "fail to reject" rather than "accept" the null hypothesis because we can never be absolutely certain in statistics — it is based on probabilities of occurrences,

TABLE 8.1
Type I and Type II Errors in Hypothesis Testing

Given that the Null Hypothesis Is:	DECISION	
	Accept H_0	Reject H_0
TRUE	CORRECT — No error	MISTAKE — Type I error
FALSE	MISTAKE — Type II error	CORRECT — No error ("power of the test")

not mathematical proof. Because we are dealing with probabilities, there is always the chance of making a mistake in our conclusions or of collecting additional data that could change the results. We measure the probability of making a mistake with alpha (α) and beta (β) values.

If the alpha value is the probability of mistakenly rejecting the null hypothesis when it is true, the probability of making the opposite error of not rejecting the null hypothesis when it is false is called beta. The beta value is derived directly from the alpha level, because the two are inversely related. If, for example, we set alpha to be 1%, which is the probability of mistakenly rejecting the null hypothesis, the probability of making the opposite error (β), of not rejecting the null hypothesis when it is true, increases. We call making an alpha error a Type I error and a beta error a Type II error, as illustrated in Table 8.1. The probability of correctly rejecting the null hypothesis when it is false is called the power of the test and is equal to $1 - \beta$.

8.3.1.1　*p* Values

Fortunately, in statistics we have a measurement, called the *p* value, that is independent of the type of statistical test we use to determine whether to reject the null hypothesis. The *p* value measures the *strength* of the results of a hypothesis test, rather than simply rejecting or not rejecting. At a given significance level (α), the *p* value is used to determine whether to reject the null hypothesis; the smaller the *p* value the more evidence there is to reject the null hypothesis.

For Six Sigma teams seeking to determine if there is a significant difference between two samples, the *p* value is the measure used to make these decisions. Table 8.2 provides accepted guidelines to determine whether to reject the null hypothesis based on the *p* value calculated from the test statistics. All statistical software and Excel provide a *p* value along with the other results of a statistical test.

In this section we have dealt with whether or not we believe, based on statistical evidence, that the results of our test are correct, regardless of the test statistic used. The next section discusses the various statistical tests and their application.

8.4　STATISTICAL TESTS AND TABLES

In this section we discuss tests of a single variable of interest and tests to determine if there are differences among two or more variables, which are known as tests of differences. Four test statistics and their distributions are presented: Z test, *t* test,

TABLE 8.2
Accepted Guidelines for Using *p* Values

p Value	Guideline
$p < .01$	Highly significant finding; reject the null hypothesis
$.01 = p = .05$	Significant finding; reject the null hypothesis at a 5% alpha level
$.05 = p = .10$	Significant finding at the 10% level; reject the null hypothesis if alpha is set at 10%
$p > .10$	Insignificant finding; little or no evidence to reject the null hypothesis

F test, and the chi-square test. Section 8.5 reviews tests to measure the strength of relationships between two variables.

The answers to the following questions determine the type of statistical test to use in analyzing sample data:

1. What is the question to be answered? Is it to determine if a process change has made a difference in the output of the process? Or do we want to know if the work output of five employees differs significantly?
2. How many variables are there? Are we interested only in testing the confidence of point estimates, or are we interested in testing whether two or more sample means are equal?
3. What is the type of data? Are the critical X's (inputs) continuous or discrete data? Are the critical Y's (outputs) continuous or discrete data?

Figure 8.2 provides a decision tree to help the team decide the appropriate test statistic.

Before we move to an example of how to use statistical tables, a discussion on the difference between *empirical distributions* and *theoretical distributions* is in order. The statistical tables we use to determine significance are theoretical distributions. When we look at a data set of numbers (e.g., a sample of the duration of 50 phone calls) we have an empirical distribution — data we have actually observed or are capable of observing. On the other hand, a theoretical distribution derives from certain basic facts and assumptions by logical mathematical reasoning, involving calculation of probabilities for many different data distributions.[5] In general, the process of inferential statistics begins with one or more empirical distributions and concludes by making reference to a theoretical probability distribution. Theoretical distributions allow us to make rational judgments concerning probability, which is the essence of inferential statistics. As discussed in Chapter 7, the normal distribution is the most common theoretical distribution. The normal PDF was worked out long ago by the statistical pioneers described in Chapter 1, such as Gauss, Poisson, and Galton. All of the statistical tables used in Six Sigma are based on theoretical distributions. They have been developed over the years to provide statistical researchers with a reference to make judgments based on their own particular empirical distributions. That is why it is important that we make an assumption on the distribution of our own empirical data — we need to be able to reference the appropriate

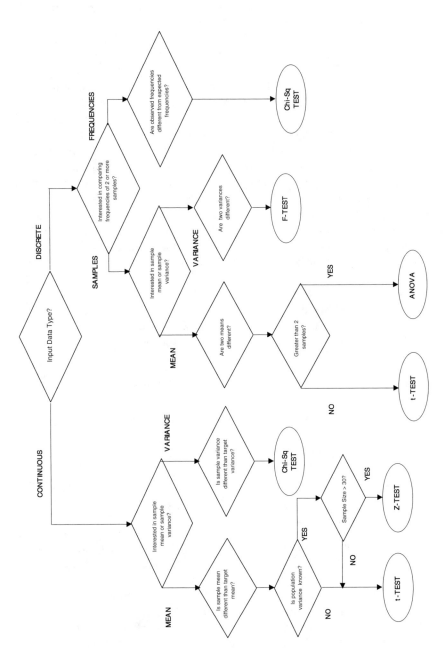

FIGURE 8.2 Choosing an appropriate statistical test.

theoretical distribution in order to make sound judgments based on the actual empirical data.

Appendix B provides the statistical tables we will use in the first example (Example 8.1) to describe the concepts and procedures of statistical tests. The example involves answering the question of whether two sample means are significantly different from each other. This is one of the more common statistical tests in Six Sigma applications because we are often interested in whether a process change has had an effect on the process outcome. In this example, taken from the case study of Card One, we are interested in whether we have improved the response time (in days) to respond to customer inquiries regarding a dispute with a merchant. The process involves entering the customer's dispute information into the Card One customer service system and contacting the merchant regarding the particular dispute. The largest single source of delay is the research conducted by the merchant, even though the merchant is contractually obligated to respond to Card One within 30 days. The customer service Six Sigma team has been working with a merchant services team to streamline communications between Card One and their merchants in order to reduce the research cycle time and the response time back to Card One, thus reducing the response time to the cardholders. Table 8.3 provides the turnaround time (in days) before the improvements and after the improvements.

TABLE 8.3
Turnaround Time (in Days) Before and After the Six Sigma Process Change

Sample 1 Before Changes	Sample 2 After Changes
27 22 23	19 24 22
27 20 23	21 21 30
27 26 25	21 17 21
22 23 24	27 22 26
19 30 23	18 19 20
$n_1 = 15$	$n_2 = 15$
$\mu_1 = 24.13$	$\mu_2 = 21.87$

Although the sample 1 mean is higher than sample 2, we cannot determine from looking at the data whether it is a statistically significant change. That is what the t test will tell us. The t test is the appropriate test because we do not know the population variance and our sample size is less than 30 observations (see Figure 8.2).

We should first look at the distribution of the data and their differences by generating a boxplot of the two data sets, as shown in Figure 8.3. Boxplots are useful to show the distribution of the data by percentiles, as well as the mean and medium of the distribution. The dot represents the mean value and the middle line is the median. The lower horizontal line is drawn at the first percentile and the top horizontal line at the third percentile. The two vertical lines (sometimes called whiskers)

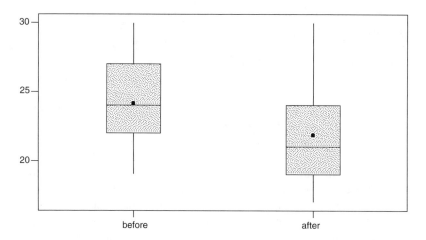

FIGURE 8.3 Boxplot of the before and after data sets.

extend to the lowest data point (at the bottom) and the highest data point. More information on interpreting boxplots can be found in the HELP section of Minitab. What is important here is that the boxplot provides a visual representation of the differences between the data and how they are distributed among the before and after data sets.

Recall from Section 8.3 that the null hypothesis states there is no difference between the sample means, or:

$$H_0: \mu_1 - \mu_2 = 0$$

$$H_1: \mu_1 - \mu_2 \neq 0$$

Thus, any difference we find between the means of the two samples should not significantly differ from zero. Other questions the team may be interested in are:

1. Is the sample 1 mean significantly greater than the sample 2 mean?
2. Is the sample 2 mean significantly greater than the sample 1 mean?

How we frame the question determines how we use the t tables found in Appendix B.2. If we are interested in determining whether one mean is higher than the other, we use a "one-tail" test of significance. If we are interested in whether the two means are equal, we use a "two-tail" test. I advocate using a two-tail test in most situations because otherwise, we are making an assumption on the behavior of the data —that one sample is larger or smaller than the other. Thus, we will use the two-tailed test of no difference for the example. This is equivalent to saying that the two samples originate from the same population source, or $\mu_1 - \mu_2 = 0$.

Other statistical tests that help determine whether two or more sample means are different are the Z test (for large samples) and the F test using analysis of variance (ANOVA) (for more than two samples; see next section). In each case, we calculate a critical value from the empirical data and compare it to the region of rejection from the theoretical distribution for that particular test value, in our example, the *t* statistic. The region of rejection is set by the alpha value we choose. Recall that alpha is the confidence or significance level. At an alpha of 5%, we are asserting that we will reject the null hypothesis of no difference between the two samples if the calculated *t* value is one that would occur only 5 times or less out of 100 (5%), or what is considered a rare chance occurrence. The region of rejection is the area that is considered a rare chance occurrence; if we get a calculated *t* value in this area, we can conclude that the chances of getting this *t* by shear chance when the null hypothesis is true are so small (5%) that the means must be different. If this is the case, we reject the null hypothesis of no difference between the means. However, if the calculated *t* is not in the region of rejection but rather in the *region of acceptance*, we say that the difference between the sample means is so small it is not significant, and we do not reject the null hypothesis, thus concluding that the two samples are from the same population source. This is equivalent to saying the process change had no significant impact on reducing the cycle time. Figure 8.4 provides test results from Minitab.

Choose Stat > Basic Statistics> 2-Sample t

```
Two-sample T for before vs. after

            N       Mean      St Dev    SE Mean
before     15       24.13      2.92       0.76
after      15       21.87      3.54       0.91

Difference = μ before – μ after
Estimate for difference:  2.27
95% CI for difference: (-0.17, 4.70)
T-Test of difference = 0 (vs. not =): T-Value = 1.91
P-Value = 0.067  DF = 27
```

FIGURE 8.4 Minitab output of a two-sample *t* test for difference of means.

The manual calculation of the critical *t* value can be found in Figure 7.22. In my experience, it is unlikely that one would need to manually calculate critical values and, thus, the focus is on interpreting the results from statistical software rather than manual calculations. Figure 8.5 provides the Excel output for the same data set.

Figure 8.6 provides a graph of the regions of rejection and acceptance for the example and where the calculated *t* value falls.We can see that in this case (Figures 8.4 and 8.5), the test results indicate no difference at the 5% alpha level, because $p = .067$, or 6.7%. This is equivalent to saying there is a 6.7% chance that we would get this calculated *t* value by shear chance if the null hypothesis of no difference were true. We can say that the results are significant at a 10% level (see Table 8.2). Although there is some evidence to support the alternative hypothesis of a difference in means, the Six Sigma team should collect more data in order to reach a firmer conclusion on the results.

t-Test: Two-Sample Assuming Unequal Variances

	Variable 1	Variable 2
Mean	24.133	21.867
Variance	8.552	12.552
Observations	15	15
Hypothesized Mean Difference	0	
df	27	
t Stat	1.91	
P(T<=t) one-tail	0.033	
t Critical one-tail	1.703	
P(T<=t) two-tail	0.067	
t Critical two-tail	2.052	

FIGURE 8.5 Excel output of two-sample t test for difference of means.

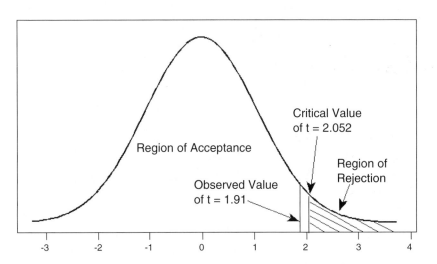

FIGURE 8.6 Regions of rejection and acceptance for the case study data, assuming $\alpha = 5\%$.

8.4.1 ANOVA

Suppose the Six Sigma team wanted to try a second process change to reduce the cycle time of responding to customer disputes, based on the results of the previous test. The data are displayed in Table 8.4. To compare the cycle times of the three processes, we conduct an ANOVA test to determine if the means of the three data sets are equal:

$$H_0: \mu_1 = \mu_2 = \mu_3$$

$$H_1: \text{At least one } \mu \text{ is different}$$

Like the two-sample test, we first construct a boxplot of the two samples to visually display the data distribution (Figure 8.7).

TABLE 8.4
Additional Data from Another Process Change for Example 8.1

Sample 1 Before Changes (Process 1)	Sample 2 After Changes (Process 2)	Sample 3 After Changes (Process 3)
27 22 23	19 24 22	18 19 21
27 20 23	21 21 30	23 20 25
27 26 25	21 17 21	23 20 22
22 23 24	27 22 26	23 22 24
19 30 23	18 19 20	18 22 20
$n_1 = 15$	$n_2 = 15$	$n_3 = 15$
$\mu_2 = 24.13$	$\mu_2 = 21.87$	$\mu_3 = 21.33$

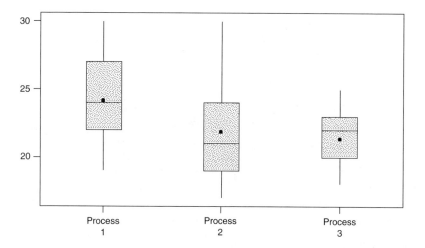

FIGURE 8.7 ANOVA boxplot for process 1, 2, and 3 data sets.

In tests concerning differences of means, such as *t* tests and ANOVA, the mathematical underpinnings are to analyze the variability *within* each sample and compare it to the variability *between* samples, similar to SPC analysis where process data are sampled. If the variability within a particular sample data set is significantly greater than the hypothesized variability between two or more sample data sets, it is logical to conclude that more homogeneity exists between the samples than within the samples and, thus, we do not reject the null hypothesis of no difference between the sample means. The within-sample variation is called the *mean square of the error* and the between-samples variation is called the *mean square of the factor*. If the ratio of the two, which is the *F* test critical value, is close to one, there is evidence the sample means are equal. The *one factor* for which we are testing differences is the mean response time, which is why the test is called *one-way* ANOVA (Figure 8.8).

Choose Stat > ANOVA > One-way

One-way ANOVA: Process 1, Process 2, Process 3

```
Analysis of Variance
Source      DF       SS       MS        F         P
Process      2    66.31    33.16     3.88     0.028
Error       42   358.80     8.54
Total       44   425.11
                                Individual 95% CI's For Mean
                                Based on Pooled St Dev
Level        N     Mean    St Dev    -------+---------+---------+---------
Process 1   15   24.133    2.924                      (---------*--------)
Process 2   15   21.867    3.543          (---------*--------)
Process 3   15   21.333    2.127     (--------*---------)
                                    -------+---------+---------+---------
Pooled St Dev =    2.923                20.8      22.4      24.0
```

FIGURE 8.8 Minitab output for one-way ANOVA results.

If we were testing for differences for both response time and response quality, we would use a two-factor ANOVA.

To calculate the F test statistic critical value, we divide the MS_{factor} (33.16) by the MS_{error} (8.54) to arrive at 3.88. We could refer to the table of critical values of the F distribution (Appendix B.3) to determine if our calculated F is in the region of acceptance or rejection. However, the p value of .028 tells us that we should reject the null hypothesis of equal means between sample data sets and accept the alternative that one or more of the sample means are not equal. In interpreting the results, it appears that the second process change, or process 3, is superior to the others, resulting not only in a slightly lower mean cycle time, but also in less variability of the data.

8.4.2 SUMMARY

Table 8.5 summarizes the application of statistical tests and their theoretical distribution tables to draw conclusions based on empirical data.

8.5 TOOLS FOR ANALYZING RELATIONSHIPS AMONG VARIABLES

In the previous section, tests of *differences* among two or more variables were presented. The statistical tools discussed in this section provides information on making decisions about whether two or more variables are *related* to each other and the strength of the relationship. The tools presented are scatter plots, correlation coefficient (r), coefficient of determination (r^2), and regression analysis.

8.5.1 SCATTER PLOTS

Scatter plots may be the most underused, yet effective tool for the Analyze Phase. Whereas boxplots help us see the distribution of the observations in a data set, scatter plots help us see the relationship between two variables. Generally, if we hypothesize that a critical Y output is dependent on a critical X input, a scatter plot will quickly help us determine the type and strength of the relationship. From there, quantitative

TABLE 8.5
Application of Statistical Tests and Tables

Table	Test	Example
Z	Test to determine if hypothesized mean equals actual population mean for samples when the population mean is known and $n > 30$	Compare actual and hypothesized values for mean time to answer a call
	Test to determine if two sample means are equal for samples when the population mean is known and $n > 30$	Determine if the mean time to answer a call is significantly different between two call centers
	Test concerning proportions when $n > 50$	Compare actual and hypothesized values of the percentage of calls answered in less than 30 sec
t	Same as Z except for cases where the population mean is unknown and $n < 30$	
F	In analysis of variance (ANOVA) tests to determine if more than two sample means are equal	Determine if mean time to answer a call varies significantly among four call centers
	Test to determine if two sample variances are equal	Determine if the variance of one call center's mean time to answer a call is different from a second call center's variance
Chi-square (χ^2)	Test to determine if hypothesized sample variance is equal to population variance	Compare actual and hypothesized values of the variance in time to answer a call
	Test to compare hypothesized frequencies to actual frequencies	Determine if the number of female call center agents is equal to the number of male agents
	Using contingency tables, determine if two or more sample frequencies vary across categories	Determine if the number of absences is different between female and male call center agents

measures of association, such as r^2 or regression can be pursued. Figure 8.9 provides interpretation of various possible relationships between X and Y.

To construct a scatter plot in either Minitab or Excel, you need two columns of data: one of the dependent (Y) values and one of the independent (X) values. Let us use another example from the case study (Example 8.2). We may hypothesize that inbound call volume is related to, or dependent on, the number of new cards issued in a given time period, because new cardholders tend to call more frequently with questions about using their card and their monthly billing statement. Instructions mailed with a new card provide an 800 number to activate the card; the same number serves as the general customer inquiry line. Figure 8.10 provides a scatter plot from two columns of data on the number of new cards issued and the number of inbound calls to the Atlanta call center in the past year. The scatter plot clearly indicates a positive relationship between new cards issued and number of inbound calls. This indicates we should further investigate to quantitatively measure the strength of the relationship.

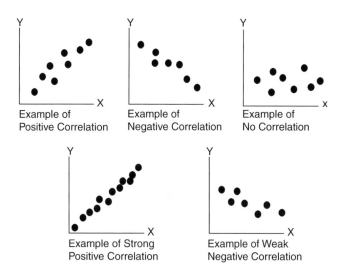

FIGURE 8.9 Examples of scatter plots indicating direction and strength of the relationship between Y and X.

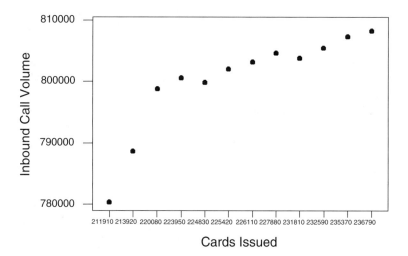

Cards Issued

FIGURE 8.10 Scatter plot of the number of inbound phone calls and new cards issued for the previous 12 months.

8.5.2 CORRELATION COEFFICIENT

The correlation coefficient, r, measures the degree to which two variables are linearly related. It varies between $+1$, indicating a perfectly positively correlated relationship, and -1, indicating a perfectly negatively correlated relationship. Values close to zero indicate no relationship exists between the two variables. Using the same data from the scatter plot example, the Minitab output (Figure 8.11) indicates an r value of .929,

Choose Stat > Basic Statistics > Correlation

Correlations: Cards Issued, Inbound Call Volume

```
Pearson correlation of Cards Issued and Inbound Call Volume = 0.929
P-Value = 0.000
```

FIGURE 8.11 Correlation coefficient for cards issued and inbound call volume.

indicating a strong relationship exists between number of new cards issued and inbound call volume. The correlation coefficient is called the Pearson correlation because Pearson is the statistician who developed the measure in the late 19th century.

The p value of .000 indicates that we should reject the null hypothesis of no correlation and accept the alternative hypothesis. Note that although we did not expressively indicate this is a hypothesis test, it is important to remember that all statistical tests are set up as a null and alternative hypothesis. That is what makes statistics a worthwhile science, because decisions using statistics are grounded in facts and proven mathematical relationships. However, a word of caution in interpreting the results of the correlation coefficient is in order. Significant correlation between two variables does not imply causation — this is an important concept that Six Sigma teams need to understand. In the example, there could be a third variable causing inbound call volume that is not related to the number of new cards issued. This is called a hidden or lurking variable. A good example is ice cream sales and drownings. The correlation coefficient between these two variables is around .91. Does this mean that ice cream sales cause drownings? Of course not. The lurking variable in this situation is the air temperature. As it gets hotter, people buy more ice cream and they also swim more. This shows that even though two variables may be *statistically* related, that does not mean they are *causally* related.

Squaring the correlation coefficient produces the coefficient of determination, or r^2. This is a measure of that portion of the variation in the dependent variable (Y) that is explained by knowing the values of X. To calculate r^2, we square the correlation coefficient (r) of .929 to arrive at .863.

If we are interested in understanding the change in Y, given a change in X, Minitab has a fitted line procedure that is found under the Regression menu. Figure 8.12 shows the Minitab output for the fitted line plot of Y and X, along with the graphical output of the plot. Because the correlation coefficient indicates a significant relationship between the two variables, we will now develop a regression equation to predict the number of inbound calls (dependent Y) based on the number of new cards issued (independent X).

8.5.3 REGRESSION ANALYSIS

Similar to correlation analysis, regression is another statistical tool for measuring relationships between variables. However, regression analysis provides further advantages because we can use it to predict values of the dependent Y, given the independent X's. Recall from Chapter 1 on the history of statistics that the definition of regression is going back to the original condition. The term was first used by Francis Galton to describe the phenomenon of how tall men's sons tended to be

Choose Stat > Regression > Fitted Line Plot

Regression Analysis: Inbound Call vs. Cards Issued

```
The regression equation is
Inbound Call = 583178 + 0.960835 Cards Issued

S = 3142.30          R-Sq = 86.2%       R-Sq(adj) = 84.9%

Analysis of Variance

Source            DF           SS           MS          F         P
Regression         1    618956313    618956313    62.6851     0.000
Residual Error    10     98740512      9874051
Total             11    717696825
```

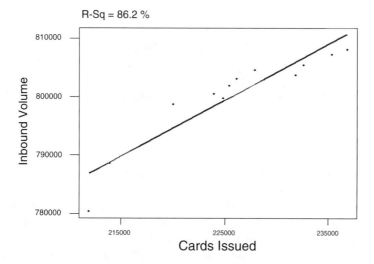

FIGURE 8.12 Fitted line plot of inbound calls vs. cards issued for Example 8.2.

shorter and how short men's sons tended to be taller — the sons' heights regressed back to the population mean.[6] Galton used the phrase "regression toward mediocrity" in his descriptions. The term r for correlation coefficient originally came from the "r" in regression.[6]

8.5.3.1 Simple Linear Regression

The type of regression analysis we will use to predict the number of inbound calls from a forecast of new cards issued is called simple linear regression. When more than one independent variable is used to predict changes in Y, it is called multiple regression, which is covered in the next section. Simple linear regression fits a straight line through the plots of X and Y data points, with the goal to find the line that best predicts Y from X. Recall from algebra that the equation of a straight line is given by $Y = mX + b$, where m is the slope of the line and b is the Y intercept. Minitab uses a similar equation with different coefficients for regression analysis:

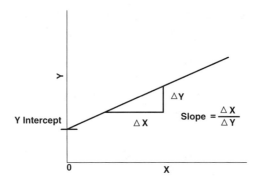

FIGURE 8.13 The intercept and slope of a regression line.

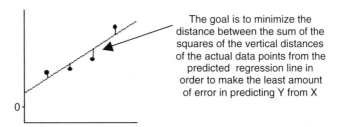

The goal is to minimize the distance between the sum of the squares of the vertical distances of the actual data points from the predicted regression line in order to make the least amount of error in predicting Y from X

FIGURE 8.14 Minimizing the sum of the squares of the distance to the regression line.

$Y = b_0 + b_1X + e$, where Y is the independent variable, X is the dependent or predictor variable, b_0 is the Y intercept, b_1 is the slope of the line, and e is an error term. The Y intercept (b_0) is the value of Y when X equals zero, defining the elevation of the line. The slope (b_1) quantifies the steepness of the line and equals the change in Y for each unit change in X. Both of these are illustrated in Figure 8.13.

To find the best means of fitting a straight line to the data, we use a method called *least squares*. The least squares method minimizes the sum of the squares of the vertical distances of the points from the line, as illustrated in Figure 8.14. This is why it is called the "least squares" method. The Minitab regression calculations and graph are provided in Figure 8.15. The first table in Figure 8.15 provides the estimated coefficients of the Y intercept (constant) and the slope (Cards Is, or cards issued), along with their standard errors (SE). The calculated *t* value tests whether the null hypothesis of the coefficient is equal to zero, and the corresponding *p* value is given. In this example, the *p* values are used to test whether the constant and slope are equal to zero. Because both *p* values are zero, there is sufficient evidence that the coefficients are not zero at a .05 alpha level, indicating a significant relationship between the two variables. The coefficient of determination (R-Sq) value of 86.2% indicates the variability in the Y variable (inbound calls) is accounted for by X (number of new cards issued).

The fitted regression line is shown in Figure 8.16. The dotted lines labeled CI are the 95% confidence limits for the inbound calls. The lines labeled PI are the

Choose Stat > Regression > Regression

Regression Analysis: Inbound Call Volume vs. Cards Issued

```
The regression equation is
Inbound Call Volume = 583178 + 0.961 Cards Issued

Predictor          Coef       SE Coef           T          P
Constant         583178         27428       21.26      0.000
Cards Is         0.9608        0.1214        7.92      0.000

S = 3142        R-Sq = 86.2%       R-Sq(adj) = 84.9%

Analysis of Variance

Source              DF            SS           MS          F          P
Regression           1     618956313    618956313      62.69      0.000
Residual Error      10      98740512      9874051
Total               11     717696825
```

FIGURE 8.15 Minitab output of the regression results for Example 8.2 data

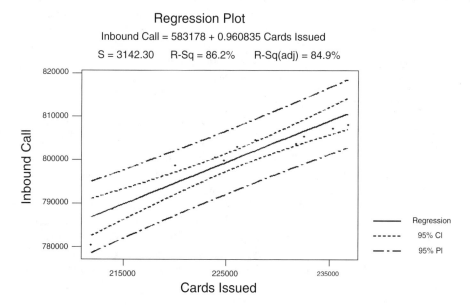

FIGURE 8.16 Regression line for Example 8.2 data.

95% prediction limits for new observations. The 95% confidence limits is the area that has a 95% chance of containing the true regression line. The 95% prediction interval is the area in which we expect 95% of all observations to fall.

The intercept value of 583,178 is the number of inbound calls expected if there were no new cards issued for a month (X = 0). The slope of 0.961 means that for each new card issued, 0.961 inbound calls are expected.

To predict the number of inbound calls using the regression equation, we need a solid forecast of the number of new cards that will be issued over the next several months. We can use the marketing department's forecast for new cards; for January and February, traditionally slow months for new card issues, the estimates are 229,500 and 227,650, respectively. Using these forecasts and the results of the regression analysis, we can predict inbound call volume at 803,727 for January and 801,950 for February, as illustration in Equations 8.1 and 8.2:

Regression Equation: Inbound Call Volume = 583,178 + 0.961 Cards Issued =

January Inbound Call Volume = 583,178 + [(0.961)(229,500)] = <u>803,727</u>

$$(8.1)$$

February Inbound Call Volume = 583,178 + [(0.961)(227,650)] = <u>801,950</u>

$$(8.2)$$

8.5.3.2 Multiple Regression

Multiple regression is similar to simple linear regression except more than one independent variable is used to predict the dependent variable. Keeping with Example 8.2, suppose that the Six Sigma team hypothesized that call volume is dependent not only on new cards issued but also on the number of times the card was used (number of charges) in a month. Together these two independent variables may provide a better prediction of inbound call volume. Figure 8.17 shows a scatter plot for the two variables. The relationship between call volume and number of charges is not as clear cut as new cards issued; however, there does appear to be some type of a relationship. Figure 8.18 provides the regression output among the variables.

The regression output suggests that adding the new independent variable (number of charges) does not significantly add value in predicting the volume of inbound calls. The t test p value of 0.794 suggests we should not reject the null hypothesis that the coefficient of number of charges equals zero, and thus we should probably not include this variable. Also, the r^2 value of 86.4% is not significantly different than the simple regression model of 86.2%, suggesting again that the addition of the new independent variable does not add to the value in the equation to predict inbound call volume. The adjusted r^2 value of 83.3% is actually lower than the adjusted r^2 value of the simple regression equation of 84.9%. This is because the adjusted r^2 penalizes the power of the prediction equation as more variables are added, because more variables tend to increase the r^2 value, without necessarily adding strength to the prediction of the dependent variable. In summary, the simple linear regression model using number of new cards issued appears sufficient to predict the inbound call volume.

8.6 TECHNICAL ZONE: SURVIVAL ANALYSIS

One issue that often occurs in Six Sigma projects is that of *censoring*. Censoring occurs when the outcome of interest in a statistical analysis is a time-dependent event, such as length of employment. Censored data are the observations in the study

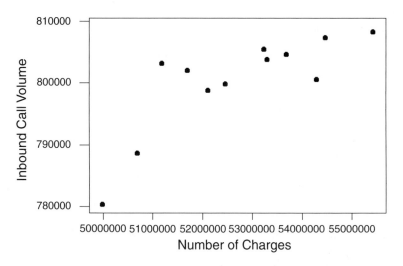

FIGURE 8.17 Number of charges and inbound phone calls for previous 12 months.

Choose Stat > Regression > Regression

Regression Analysis: Inbound Call vs. Cards Issued, Number of Ch

```
The regression equation is
Inbound Call Volume = 579378 + 0.909 Cards Issued + 0.00030 Number
of Charges
```

Predictor	Coef	SE Coef	T	P	VIF
Constant	579378	32073	18.06	0.000	
Cards Is	0.9085	0.2324	3.91	0.004	3.3
Number o	0.000296	0.001101	0.27	0.794	3.3

```
S = 3299        R-Sq = 86.4%      R-Sq(adj) = 83.3%
```

Analysis of Variance

Source	DF	SS	MS	F	P
Regression	2	619744069	309872035	28.47	0.000
Residual Error	9	97952756	10883640		
Total	11	717696825			

FIGURE 8.18 Minitab output of the multiple regression results for Example 8.2 data

that have NOT experienced the time-dependent events that are of interest in survival analysis studies. Survival analysis is a group of statistical methods for analyzing and interpreting survival data. The opposite of survival is failure; in manufacturing, a critical area of study is estimating product failures, particularly mean time to failure. Estimating the mean time to failure for a product or process is called reliability analysis.

Suppose we are interested in reducing the turnover of telemarketing associates and the critical Y outcome is length of employment. We hypothesize that years of experience and whether the associate has a college degree influence the length of employment. However, if we randomly sampled personnel files of associates to collect data on these variables, our random sample would likely contain associates who are still employed at the time of the study (i.e., who have not experienced the time-dependent event of leaving the company). Why can't we just sample the agents who have already left? The issue is bias — there may be some characteristic inherent in those who left vs. those who stayed that could affect the outcome of interest. For example, associates with more years of experience at the time of hire may stay longer and, thus, just sampling those who left will lead to biased results. Many complex and sophisticated statistical tools and techniques for survival analysis can be used that help estimate parameters associated with reliability, such as mean time to failure.[7]

The problem of censored data in servicing environments, such as estimating length of employment at hire date, requires both common sense and analytical approaches. Suppose the average tenure of an associate is 12 months, with a standard deviation of 3.5 months. By choosing to randomly sample only the personnel files with a hire data older than 19 months, we have some confidence in the validity of the research. This is because 12 months plus 2×3.5 months = 19 months ensures that about 95% of the files sampled will not contain censored data. Another method may be to randomly sample current data and analyze what variables are significantly different between those agents who have left and those still employed.

8.7 SUMMARY

The Six Sigma team should exit the Analyze Phase with a clear understanding of the root causes of the problems associated with transforming the critical X's into critical Y's. Identification of the critical X's starts in the Measure Phase through various measure tools and techniques. The hypothesized critical X's are then either confirmed or discarded in the Analyze Phase using quantitative statistical techniques. The Improve Phase focuses on improving the critical X inputs and the processes that transform the inputs into outputs.

It is unlikely that all the tools in the Measure and Analyze Phases are used to confirm the root causes of the problems. The team may even use tools that have not been covered in these two chapters. The objective is to provide the Six Sigma team with the most frequently used tools and techniques in improving transactional processes in a service environment.

Pulling together and preparing a concise presentation of all the findings of the Measure and Analyze Phases are no small tasks. However, it is highly recommended that the team do just that at this point, before moving on to the Improve Phase. It is important to have a tollgate meeting at the end of the Analyze Phase so that the project sponsor and Champion understand what the team will focus on during the Improve Phase. Summarizing the measurement and analysis findings at the end of the Analyze Phase will also provide the team with clearer direction on precisely what root cause problems need to be solved in the Improve Phase.

REFERENCES

1. Shapiro, B. et al., Staple yourself to an order, *Harvard Business Review*, July 1992, pp. 113–122.
2. Goldratt, E.M. and Cox, J., *The Goal: A Process of Ongoing Improvement*, North River Press, Great Barrington, MA, 1992.
3. Dettmar, W., *The Theory of Constraints*, ASQ Quality Press, Milwaukee, WI, 1992, p. 17.
4. Goldratt, E.M. and Cox, J., *The Goal: A Process of Ongoing Improvement*, North River Press, Great Barrington, MA, 1992, p. 60.
5. The theoretical distribution is known as the sampling distribution, which is different from the sample (empirical) distribution. The sampling distribution is the distribution that results from drawing many different samples from a specified population — each sample has it is own mean and standard deviation, thus forming a distribution or particular shape of its own. This resulting shape is what is called the theoretical distribution and the statistical tables are based on this. The sampling distribution is also what makes the central limit theorem work.
6. Zikmund, W.G., *Business Research Methods*, 4th ed., Dryden Press, New York, 1994, p. 557.
7. I refer the interested reader to the following text, which has an excellent discussion of reliability analysis in a Six Sigma context: Breyfolge, F., III, *Implementing Six Sigma: Smarter Solutions Using Statistical Methods*, John Wiley & Sons, New York, 1999.

9 Improve Phase

9.1 OVERVIEW

The Six Sigma team should start the Improve Phase with a clear understanding of the root causes of the problems associated with transforming the critical X's into critical Y's. The Improve Phase focuses on improving the processes that transform the inputs into outputs. The primary result of the Improve Phase is a redesigned process incorporating all the lessons learned through the project. There should be clear evidence that solutions generated and integrated into redesigned processes are capable of closing the gaps between the current process and the customer's CTQ requirements. The new process should also demonstrate direct cost savings or increased revenue to the company, given that Six Sigma is a business strategy to increase bottom line profits. Regardless of the objective of the project, bottom line profitability is a key outcome. For some projects, direct cost savings may be difficult to quantify; techniques for estimating the monetary impact of projects are covered in this chapter.

The Improve Phase is when we generate alternative process designs and then analytically choose the one that will best meet the project objectives. The material in this chapter, which is outlined below, will assist the team in this process:

1. Process redesign principles
2. Development of alternative solutions
3. Design of experiments
4. Pilot experiments
5. Cost/benefit analysis
6. Implementation plan

The Improve Phase may be the most interesting part of a Six Sigma project. The opportunity to develop and test creative solutions is the payoff for all the hard work accomplished in prior phases. It can be difficult to withhold offering solutions to problems in prior phases, but Six Sigma is an approach of scientific discovery in which the facts are assembled into a coherent picture in the first three phases, and only then can we apply our knowledge to develop an optimal solution.

9.2 PROCESS REDESIGN

In prior phases, the CTQ input measures and the process measures have been identified and their baseline performance determined. We have matched the baseline process performance against the customer and business CTQs and identified the size of the gap that needs to be closed in order to meet the CTQs. The Improve Phase is about redesigning the processes to close the gap.

Many Six Sigma projects will have several redesign opportunities, because subprocesses feed into the overall process. For example, in our case study to reduce call volume into the Card One customer service centers, many different subprocesses, such as interactive voice response, routing, staffing, and scheduling, feed into the overall process of delivering customer service through a centralized 800 number.

The first step in process redesign is planning. This involves clearly stating the objectives of the new process in operationally defined terms. Is it to achieve a 20% reduction in cycle time or is it to increase the percentage of calls answered in less than 30 sec to 80% from the current 50%? The project charter should be revisited at this point to ensure that the project is on track to meet the objectives originally defined in the charter. Next, the Six Sigma team is ready to translate the required targets of the redesigned process into a vision of the new process. Start with a "theoretical pie in the sky" if necessary and work toward the a realistic goal given the business and cost constraints.

9.2.1 Process Redesign Principles

Recall from the previous chapter that in analyzing process maps, certain process design principles have stood the test of time through many process and quality improvement initiatives. Adherence to these principles ensures the redesigned process is optimal and capable of meeting the customer and business CTQs:

1. *Minimize approvals, especially multiple levels of approvals.* Design the process so that controls are built in rather than externally applied by a supervisor or other employee. For example, rather than routing purchase orders through an approval process, design a system where employees can order only what has been previously approved for that position. If an employee in a particular position, such as a business analyst, needs a certain type of desktop computer configuration, the system should present only the preapproved selection to him or her. The same is true for office and other supplies. Build in process and system controls so that external and manual control efforts are minimized or eliminated completely.

2. *Minimize hand-offs* because hand-offs create queues as well as disconnects in the workflow, leading to defects and long cycle times. Hand-offs may be one of the most damaging characteristics in process flows. Unfortunately, this is a widespread problem because work typically flows horizontally across departments even though control flows vertically within functions. Not only do departmental hand-offs create increased opportunities for defects and potential for losing whatever is flowing through the process (such as orders), they do not encourage taking responsibility for the entire process. Customers find themselves in a maze of trying to track down responsibility for their order while the different departments conveniently point fingers at each other. If feasible, reorganize around work

processes rather than functional units. This principle applies at all levels of the organization, from the sub-subprocesses within a small area of a department to the entire corporation.

3. Similar to #2, *assign clear accountability for process steps*. If everyone is responsible, no one is responsible. This usually happens when a poorly defined "work team" is assigned responsibility. If one team member hands off work to a second team member, and there is ultimately a process failure, it is finger pointing time again. Redesign the process so that accountability is built in. If low-level clerical workers handle an entire sub-subprocess, make them accountable for the quality of that process. Build the responsibility into their compensation so they are motivated to take ownership of their actions. A related redesign principle is to push responsibility for work down to the lowest level possible, along with the training and authority to get it done. This is especially important for employees who come into direct contact with the company's external customers. Common sense dictates that a well-trained customer service representative who has the authority to do the "right" thing for the customer at the moment of truth will result in increased customer satisfaction. Processes should be designed so that the "right" actions are agreed to beforehand, such as specifying dollar limit approvals in granting immediate customer refunds, to ensure process control.

4. *Build quality control into each process step* to minimize defects at all steps along the process. This allows defects to be corrected immediately rather than promulgated throughout the entire process. It is surprising how this one process design principle can so significantly decrease the costs of a process. However, if we think of the opposite situation, in which quality control is applied at the end of the process, we quickly grasp the implications — defects early in the process are allowed to move downstream through the process only to be found at the end, after many other costly process steps have been performed on the defective product. A related design principle is to incorporate error-proofing activities in every process step to avoid producing defects or to identify them early in order to take action to prevent them in the future. This is covered in more detail later in Chapter 12 on Poka-Yoke, or error proofing, techniques.

5. Closely related to #4, *minimize or eliminate inspections or appraisal activities*. Inspections add unnecessary time to the process, yet they do not guarantee accuracy. Studies have shown that only 80 to 95% accuracy is achieved through inspection.[1] You cannot inspect quality into a process or a product.

6. *Balance the flow of the work to avoid bottlenecks*, because bottlenecks create queues and queues create inventory and increased cycle time. Later in this section is a discussion of the theory of constraints, a process improvement approach that states that the total output of a process is limited to its slowest process step. If we design a process to continuously flow by balancing the volume of work and time through each process step, queues and work in process are minimized. This leads to increased

throughput in a shorter amount of time at less costs, while increasing or maintaining the quality of the output. In a transactional environment, queues are analogous to work-in-process inventory in a manufacturing environment. High levels of inventory, whether it is customer orders waiting to be processed or electronic subcomponents waiting for assembly, increase costs and cycle time. This, in turn, leads to decreased customer satisfaction, and the process can quickly spiral out of control.

7. To support the design principle of balanced, continuous workflows described in #6, *minimize batch sizes* of the work flowing through the process. This will reduce the cost of goods sold and cost of quality as well as shorten delivery schedules. Small-batch workflow is discussed at length in Chapter 12, Section 12.4.

8. *Design the process to handle the routine rather than the exceptional.* Michael Hammer was the first to point out that many business processes are "overengineered" or designed for the worst-case scenario.[2] Most likely, the team will discover existing process steps that were developed to handle a single exception that no longer applies. If exceptions occur, handle them off line. Design the process according to the Pareto principle — build the 20% of the process that will handle 80% of the cases. Actually, it will probably be along the lines of 90/10, because if "exceptions" account for more than 10% of the situations, significant effort will need to be directed at understanding the exceptions and taking steps to eliminate them. Create specific backup procedures to handle the 10% or less of the exceptions. Consider a healthcare Six Sigma team trying to redesign a patient admission process. For the small percentage of patients (veterans) who require a different admission procedure, one must enter data into a special government information system. There is no need to provide system training and access to all admissions clerks; simply train one or two clerks plus the supervisors, so that veterans will have the same level of service as everyone else. The extra time and dollars spent making the exception part of every clerk's routine process is probably not worth it. In summary, it is usually better to build a simple process for most of the situations rather than a robust yet complicated one that will handle every situation.

9. *Question everything.* Process redesign is the time for creativity. Can the work be outsourced? Can it be handled by another employee at the original source of the data? Why is the work done in this department by this position? Can the work be done at other, more advantageous times? Can process steps be combined, rearranged, or even reversed? Use the principle of "touch it — work it," which minimizes the number of times a work product is handled. Always consider the perspective of the customer, whether internal or external, as the new process is developed.

9.2.2 Automation Considerations

Often, a Six Sigma team will be faced with the choice of automating a previously manual process. There are many benefits to automation, especially for routine, mundane tasks. However, a word of caution is in order. Do not automate a previously manual process without first redesigning the manual process for optimal performance. Otherwise, the team is just taking a poorly designed process and making it run faster and perhaps more expensively, without understanding the root cause of the performance issues. Instead, redesign the manual process to perform optimally and then take steps to automate the optimized process. This is especially true for Design for Six Sigma projects, which are discussed further in Chapter 11.

9.3 GENERATING IMPROVEMENT ALTERNATIVES

At this point in a Six Sigma project, the team probably has already begun to think about solutions for the root cause problems. However, this is the appropriate time to formally develop creative solutions based on generating alternatives. The tools covered in this section are brainstorming, affinity diagrams, and force field analysis.

9.3.1 BRAINSTORMING

Brainstorming is a technique designed to produce a lot of ideas quickly. It is not meant to determine the best solution to the problem — that will come later. A number of creative techniques exist for brainstorming. With the "popcorn" method, all team members in the session spontaneously call out ideas and a facilitator writes them on a flip chart. Another technique is the "silent" method, where team members silently write down all of their ideas on Post-It notes and then the notes are collected and shared with the entire team. For the "anti-solution" technique, the team is asked to think of all the ways they could make the process *worse* and the customers *more dissatisfied*. The results are then turned around in search of the best solutions.

Unlike other Six Sigma techniques, brainstorming favors quantity over quality. The idea is not to generate the best solutions but to generate a lot of far-reaching solutions. Usually teams like to set ground rules at the beginning of a brainstorming session in order to obtain the maximum number of ideas. Following is a standard list of brainstorming guidelines that the team may customize for their own purpose:

1. There are no dumb ideas. Do not suppress or comment on the ideas of others because this may inhibit everyone from expressing their ideas. This is a creative process — don't get too serious.
2. Team members should be encouraged to build on others' ideas. Because this is a spontaneous, creative process, one person's thoughts are likely to generate another's thoughts along the same idea.
3. Choose the facilitator carefully. Facilitators should be open-minded and extroverted. They must be able to write fast if using the popcorn method. Make sure the appropriate materials are in the room, such as flip charts, markers, and tape.

These guidelines are for "structured" brainstorming; but there is no reason not to have many unstructured brainstorming sessions within the Six Sigma team at any time in the DMAIC process. You may want to keep an ongoing list of ideas generated by the team during project meetings to be consolidated and utilized at the appropriate time.

9.3.2 AFFINITY DIAGRAMS

Affinity diagramming often follows a brainstorming session, with the goal of providing more structure to the many ideas generated. The objective of affinity diagrams is to develop meaningful ideas from a list of many unrelated ideas. It is helpful when assorted random ideas need to be clarified in order to reach team consensus or when a large, complicated issue needs to be broken down into smaller categories. An affinity diagram may also help with understanding the pros and cons of a potential improvement solution, especially if it is likely to be controversial. The objective is to understand all facets of the risks associated with the potential solution, including issues involving acceptance, implementation, and likely results.

Affinity diagrams are typically developed in brainstorming-like sessions, where the objective of the effort is clearly defined and the participants quickly generate ideas that will be grouped by category. Team members may write their ideas on Post-It notes and the facilitator collects the notes for display. After the note contents are reviewed, the team agrees on category headings. Team members then take about 10 to 15 min to place each note under a particular category, setting aside any duplicate ideas. Category headings are not planned in advance but evolve based on the group's ideas. The team develops the affinity diagram based on the category headings and the issues listed under each category. From there the team can discuss the particular issues associated with each category and take steps to further explore the issues in a more structured approach.

9.3.3 FORCE FIELD ANALYSIS

Force field analysis is another brainstorming-like technique that helps the team understand the driving and restraining forces associated with change. A facilitator draws two columns on a flip chart and labels them driving forces and restraining forces. Driving forces are those issues or actions that push for change whereas restraining forces stand in the way of change. After the team has brainstormed both the driving and restraining forces, they analyze the list to determine if one side has a clear advantage over the other. Change is unlikely if no side has a clear advantage. If the restraining forces are greater than the driving forces, the team should further brainstorm ideas and strategies to overcome the restraining forces to bring about change. Figure 2.1 illustrates a force field analysis on the driving and restraining forces associated with implementing a Six Sigma initiative at Card One, the company from our case study.

9.4 TECHNICAL ZONE: DESIGN OF EXPERIMENTS

It is safe to say that statistical design of experiments, called DOE, is utilized more in manufacturing than transactional servicing. Widespread adoption of this tool has not yet occurred in most service-related operations, such as financial services call centers, hospitals, and retail outlets, most likely because it is perceived as a complex tool. This is unfortunate, because DOE has the potential to provide significant benefits in a complex service environment, where there are many critical X's influencing the critical Y's. As with most statistical tools, the sense of fear and mystery associated with DOE is caused by lack of clear explanations and examples in most training courses and textbooks. I have attempted to provide a clearer presentation of DOE concepts, application, and benefits in this section, to promote greater use of the tool in nontraditional situations.

DOE is a method to understand how the critical X inputs into a process affect the process outcome, or critical Y. DOE assists the Black Belt in determining the vital few X's that have the greatest effect on the critical Y outcome, and the optimal levels or settings of those vital few X's. Although regression analysis is used for similar purposes, the advantage of using DOE is that the effects of the critical X's interacting with each other, as well as their individual effects on the outcome, can be determined. Additionally, different levels of the X's can be tested interactively and individually to determine the impact on the response. Situations in which DOE is useful include:

- Choosing between alternatives, such as alternative suppliers or accepting or rejecting a job candidate or credit card applicant. This is called a comparative design.
- Narrowing down the list of potential critical X's to the vital few critical X's that have the most significant impact on the process outcome. This is called a screening design because it screens for the vital few X's.
- Determining if interactions among the vital few critical X inputs make a significant difference to the outcome.
- Determining the level of the critical X that provides the optimal critical Y output. This is an optimizing design that leads to development of a regression equation to predict the critical Y outcome, based on the critical X's identified in the experiment.

To illustrate the concepts and application of DOE, we introduce Example 9.1 from the Card One case study. Suppose a Six Sigma team at Card One is tasked with increasing sales in telemarketing. The Card One telemarketing department solicits new card members to offer a variety of financial products. The human resources manager suspects that the more experienced agents have higher sales, but previous data analysis is inconclusive. The telemarketing manager thinks that younger agents, with less experience, seem to do better. The company has also been hiring more college graduates in recent years and the human resources manager would like to determine if this has increased sales. The Black Belt decides to perform a DOE to test the theory that the combination of years of experience, college degree, and age make a difference in sales. The first experiment is a screening design that will help screen the vital few critical X's that have the most impact on the outcome of sales. Then a regression model will be developed to predict sales based on the vital few critical X values.

The team could carry out the experiment one factor (X) at a time (known as OFAT). OFAT is equivalent to performing three different experiments to predict the outcome of sales, holding two input variables steady while manipulating the third. Not only is this method time consuming, but the potential interaction effects among the independent variables are not measured. For example, there may be an interaction between years of experience and age. However, if we measure the effect of these two factors on the outcome by themselves, their interaction effect would be lost.

As with most experiments, careful planning is required upfront to ensure the results of the experiment satisfy the objectives, with effective use of time and resources. The steps involved in performing DOE are:

1. Set the objectives of the experiment. This involves:
    *   Selecting the critical Y outcome (response) and critical X input (independent) variables.
    *   Setting the levels of the independent variables.
2. Select an experimental design.
3. Set up the design using a statistical software package such as Minitab. (Note: You may also find DOE tables in many statistical books.)
4. Execute the design. This involves actually collecting the data and entering the data in the design table (Minitab worksheet).
5. Interpret and take action on the results.

Figure 9.1 provides a diagram for the example, illustrating the input and output variables of interest.

DOE has its own terminology, as described below.

Factors: The inputs into the model (critical X's) that drive the outcome. In DOE, the factors must be discrete; so if the input is naturally continuous, such as driving distance, we must code the data into two or more levels.

Levels: The values that a factor can take on, classified as two or more categories, such as low and high (two levels) or low, medium, and high (three levels). The different values of each level need to be far enough apart so that the effect on Y is appropriately measured.

Factorial design: The number of factors (denoted by k) and levels in the experiment. It is expressed as $k1 \times k2 \times k3 \dots kN$. A $2 \times 2 \times 3$ design indicates an experimental design with three factors. The first factor has two levels, the second factor has two levels, and the third factor has three levels. The shorthand notation is (levels)^{factors}, so a 2×2 design = 2^2 and a $2 \times 2 \times 2$ design = 2^3.

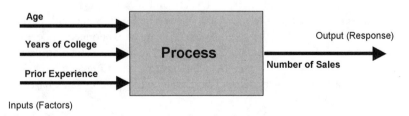

FIGURE 9.1 Design of experiment process model for Example 9.1.

Test run (also called trial): A single combination of factors and levels that yield one response variable.

Treatment: A single level assigned to a single factor during an experimental run, such as driving distance set at less than 15 mi. A treatment combination is an experimental run using a set of specific levels for each factor.

Replication: Running the same treatment combination more than once. Note: this does not mean that a particular treatment combination is run sequentially, but that the entire experiment is run again. Replication provides more confidence to the test results.

Main effect: The effect on the outcome (Y) observed from changing the setting of a single factor.

Interaction: The combined effect on the outcome of two or more factors working together. It occurs when one factor's effect on an outcome depends on the level of another factor.

Randomization: A schedule for running treatment combinations in a random manner (i.e., the conditions in one run do not depend on the conditions in a previous run).

In Example 9.1, we will determine the inputs and outcome and use Minitab to set up, execute, and interpret the DOE. In general, we are looking at a three-step process in Minitab: (1) setting up the design table in a worksheet by running the DOE design function, (2) entering data collected from the experiment into the worksheet, and (3) obtaining results from the experiment by running the DOE analyze function.

EXAMPLE 9.1

1. Set the objectives of the experiment. The objective of this experiment is to increase sales through improved hiring decisions of telemarketing recruits. The hypothesis is that age, years of college, and prior telemarketing experience all affect sales. The Y response variable is sales and the input factors (X's) are age, years of college, and prior telemarketing experience.

2. The experimental design is a **2 × 2 × 2** or 2^3 design. The levels for each factor are described below:

Factor Name	Units of Measure	Low Level (–)	High Level (+)
Age	Years	20	40
Years of college	Years: 1 = less than 2 years; 4 = 2 years or more	1	4
Prior telemarketing experience	Years: 0 = less than 1 year; 1 = 1 year or more	0	1

3. The Minitab output from running the DOE design function for the 2^3 factorial table is provided below:

Choose Stat > DOE > Factorial > Create Factorial Design

Run Order	Age	Years of College	Prior Experience
1	40	1	0
2	40	1	1
3	40	4	0
4	20	4	0
5	20	1	0
6	40	4	1
7	20	4	0
8	20	4	1
9	40	1	1
10	40	4	1
11	20	1	1
12	20	1	0
13	40	4	0
14	40	1	0
15	20	4	1
16	20	1	1

4. Execute the design. This involves collecting the data and entering them in the design table (Minitab worksheet). The data can be actively collected or historical data can be used. The results of the experiment are provided in the table below. The sales figures are entered into the Minitab worksheet under a new column heading called "sales."

Choose Stat > DOE > Factorial > Create Factorial Design

Age	Years of College	Prior Experience	Sales
40	1	0	86
40	1	1	86
40	4	0	133
20	4	0	126
20	1	0	85
40	4	1	70
20	4	0	139
20	4	1	88
40	1	1	69
40	4	1	86
20	1	1	71
20	1	0	111
40	4	0	140
40	1	0	116
20	4	1	68
20	1	1	82

Choose Stat > DOE > Factorial > Analyze Factorial Design Fractional
Factorial Fit: Sales vs. Age, Years of Col, Prior Experience

Estimated Effects and Coefficients for Sales (coded units)

Term	Effect	Coef	SE Coef	T	P
Constant		98.13	3.721	26.37	0.000
Age	3.00	1.50	3.721	0.40	0.697
Years of	16.50	8.25	3.721	2.22	0.057
Prior Ex	-41.00	-20.50	3.721	-5.51	0.001
Age*Years of	-1.25	-0.63	3.721	-0.17	0.871
Age*Prior Ex	-2.75	-1.38	3.721	-0.37	0.721
Years of*Prior Ex	-15.75	-7.87	3.721	-2.12	0.067
Age*Years of*Prior Ex	1.00	0.50	3.721	0.13	0.896

Analysis of Variance for Sales (coded units)

Source	DF	Seq SS	Adj SS	Adj MS	F	P
Main Effects	3	7849.0	7849.00	2616.33	11.81	0.003
2-Way Interactions	3	1028.7	1028.75	342.92	1.55	0.276
3-Way Interactions	1	4.0	4.00	4.00	0.02	0.896
Residual Error	8	1772.0	1772.00	221.50		
Pure Error	8	1772.0	1772.00	221.50		
Total	15	10653.8				

FIGURE 9.2 Minitab output from the DOE screening design.

5. Interpret and take action on the results. After entering all the results of the experiment into the new column, it is time to analyze the results of the experiment (Figure 9.2).

The p values in the top table of Figure 9.2 determine which of the effects are significant. With an alpha level of .10, the main effects for years of college and prior experience and the years of college*prior experience interaction are significant. Next, we will repeat the DOE analyze function using only the terms identified as significant ($p < .10$), screening out the insignificant variables of age, the two-way interactions of age*years of college and age*prior experience, and the three-way interaction of age*years of college*prior experience (Figure 9.3).

Because all the factors in the sales model have p values less than .05, we can be assured that the model is a good fit to the data. The next step is to generate main effects plots (Figure 9.4). Note the following from Figure 9.4:

- Years of college is the difference between the low setting and the high setting on the graph. This means that employees with more than 2 years of college generate higher sales than those with less college.
- Prior experience is the difference between the two categories. This means employees with less than one year's experience generate higher sales compared to those with more than one year's experience.

We see that prior experience has a bigger main effect than years of college. That is, the line connecting the mean responses for less than 1 year of experience and

Choose Stat > DOE > Factorial > Analyze Factorial Design Fractional Factorial Fit: Sales vs. Years of Col, Prior Experience

```
Estimated Effects and Coefficients for Sales (coded units)

Term                   Effect      Coef    SE Coef       T       P
Constant                          98.13      3.103   31.62   0.000
Years of                16.50      8.25      3.103    2.66   0.021
Prior Ex               -41.00    -20.50      3.103   -6.61   0.000
Years of*Prior Ex      -15.75     -7.88      3.103   -2.54   0.026


Analysis of Variance for Sales (coded units)

Source               DF      Seq SS      Adj SS     Adj MS      F       P
Main Effects          2      7813.0     7813.00     3906.5   25.36
0.000
2-Way Interactions    1       992.3      992.25      992.3    6.44   0.026
Residual Error       12      1848.5     1848.50      154.0
  Pure Error         12      1848.5     1848.50      154.0
Total                15     10653.7
```

FIGURE 9.3 Minitab output from fitting the sales model of significant factors.

Choose Stat > DOE > Factorial > Factorial Plots

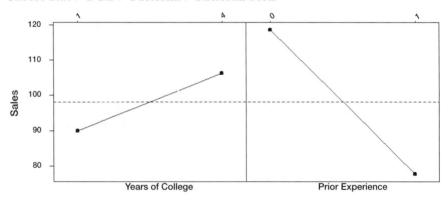

FIGURE 9.4 Main effects plots for the sales model: one for years of college and one for prior experience.

greater than 1 year of experience has a steeper slope than the line connecting the mean responses at the low and high settings of years of college. This says that employees with less telemarketing experience had higher sales, which is counter-intuitive. The telemarketing manager is not surprised, though, because the nature of the scripts for the telemarketers requires strict adherence in order to maximize sales. The company invested a significant amount of money developing the scripts, bringing in a psychologist, a linguist, and a well-known author of effective sales scripts.

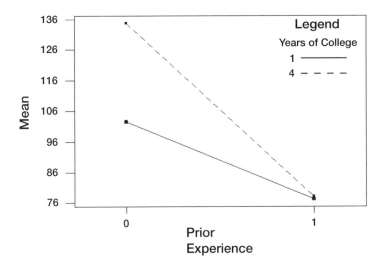

FIGURE 9.5 Interaction plot for prior experience and years of college.

Telemarketers with more experience tended to have their own approach to the sale, which was not as effective as relying solely on the specialized script.

Although experience appears to affect sales more so than years of college, it is important to look at the interaction. An interaction can magnify or cancel out a main effect. The next step, then, is to look at the significant interaction between prior experience and years of college, illustrated in Figure 9.5. Although we have already verified a significant interaction with the model fitting results, the interaction plot visually tells us the degree of this effect. An interaction plot shows the impact that changing the settings of one factor has on another factor. Because an interaction can magnify or diminish main effects, evaluating their effect is extremely important. The significant interaction between years of college and experience shows up as two lines with sharply differing slopes.

Telemarketers with little prior experience have greater sales at both levels of college. However, we see that the difference in sales between those with little prior experience and more college is much greater than the difference in sales between those with little prior experience and little or no college.

Thus, to get the highest sales, the telemarketing manager should favor hiring people with more college and little or no experience. Of course, other hiring considerations would include enthusiasm, flexibility, and interview performance.

To summarize, the following steps were taken in using DOE to help with hiring decisions for telemarketers:

1. We decided on a design for the experiment, the critical X's and Y, then generated and saved settings using Minitab's Create Factorial Design command.
2. We ran the experiment and entered the responses.

3. We examined the output and screened out the unimportant X's of age and the interaction effects of age with the other critical input variables of years of college and prior experience.

4. We generated main effects and interaction plots with Minitab's Factorial Plots command to visualize the effects of the critical X's on the output of sales.

5. We learned from our analysis that to maximize sales, we should look to hire telemarketers with more than 2 years of college and less than 1 year of experience, among other considerations.

9.5 PILOT EXPERIMENTS

Similar to statistical DOE are pilot experiments. Typically, this involves keeping a "control" group the same as the existing process and implementing a "pilot" group under the new process. Careful planning must be done upfront to ensure that time and effort are well spent. The primary considerations for a pilot experiment include (1) the objectives, (2) the sample characteristics, (3) the timeframe, and (4) the tracking and interpretation of the results.

The first step must be to state the hypothesis of the pilot, much like stating a statistical test hypothesis. Pilot experiments are often called champion–challenger. The champion is the defender of the status quo (current process), which could be considered the null hypothesis. The challenger "challenges" the status quo with a better or "alternative" idea.

The structure of the experiment depends on what is being tested. If the objective is to test whether a new process produces better performance, the operational definition of "better performance" needs to be spelled out clearly in the experiment plan. One must also consider how the pilot will affect the customer, especially the external customer, whom the Six Sigma team cannot typically bring into the planning process. If the objective is to determine whether a new process has a significant effect on improving internal efficiency and effectiveness measures, with little or no impact to the customer, then the experiment is somewhat easier to manage.

Similar to statistical DOE, an experiment design plan should be developed and agreed upon prior to execution of the experiment. Listed below are the suggested sections of a design plan, along with descriptions:

1. Overview: This section provides a general description of how the experiment fits into the larger Six Sigma project. Is it testing one change out of many in the project recommendations, or does the experiment comprise the major output of the project? The scope of the planning and execution may be different if the entire project success depends on the one pilot rather than if it involves one recommendation out of many in the project. The overview needs to include the background of how the recommendation that the experiment is testing evolved and, similar to the project

charter, state the business case (why the pilot should be done) and the general goals of the pilot experiment.

2. Hypothesis: What exactly is being tested and what answers will the experiment answer? How will the team know if the pilot is successful? What are the key outcome measures, stated in terms of operational definitions? How will they be tracked? For example, if a Six Sigma team wants to determine the effectiveness of a new collections script against the existing script, the hypothesis may be that the new collections script will increase the performance of the collectors. However, the measure of "increased performance" needs to be operationally defined as, say, increased dollars collected per hour, keeping all the other performance measures such as right party contacts and promises to pay constant. Or one could include all three categories. The point, of course, is that the hypothesis of the study needs to be clearly stated in the experiment plan. The tracking of the pilot results needs to mirror the hypothesis. It certainly helps if there is a target that can be set beforehand, such as "dollars collected per hour will increase at least 20% through this new script." Often, however, the team does not have a target and the objective will be to determine if "dollars collected per hour is statistically significantly higher" under the new script.

3. Sample: What are the data sample characteristics (i.e., who, how many, and when)? To minimize disruptions in the normal departmental workflow, the sample may be limited to the minimum number of data observations that ensure statistically valid results. Other teams may be more confident of the experiment outcome and split the pilot group with the control group 50–50. If the team is not sure, it is usually best to go with the smaller scale, statistically significant sample, and then, depending on initial results, expand to a wider sample until the entire population is operating under the new process. This often happens in clinical trials when testing the efficacy of a new medical treatment or drug. If the results of the new treatment prove to be significantly better than the control group, and the pilot participants' health is at stake, the experiment is ended early and the new treatment is provided to everyone.

   The size and characteristics of the sample will differ according to each situation. If, in the example above, we are interested in whether the new script is more effective in dollars collected per hour, the one-sided two-sample t test is probably the appropriate measure to determine if the pilot results are significantly higher than a control group under the existing script. There are several choices for a control group, depending on the other influences on dollars collected. It could be as simple as having a randomly selected group of collectors use the new script during the same time period that the rest of the department collectors use the existing one. Another method may be to use the same group of collectors and use the old script for a period of time, say a week, and the new script the next week. Still another is to have the same collectors use both scripts during the same period, randomly alternating between the two scripts. After enough sample observations have been documented, results are compared, using a two-sample *t* test with a one-sided hypothesis. Depending on the results, additional statistical analysis may be in order, such as analyzing the results by collector or by debt amount. It is important to remember that in choosing the sample size for the random sample of accounts to be tested, the sample size will need to be larger to have the same margin of error as a test of just differences. If the experiment involves determining if the pilot data are either better or worse than the control data — not just different — then, the rule of thumb is to multiply the margin of error for a simple test of differences by 1.7 to get the margin of error for a one-sided, better-or-worse test.[3] The way to maintain an acceptable margin of error for the one-sided test is to increase the pilot and control sample sizes.

4. Timeframe: The timeframe to run the experiment and measure the results differs depending on what is piloted and the experiment objectives. Seasonality of the data needs to be considered as well as the time it takes for the experiment results to be fully realized. For example, in testing the effectiveness of different collection scripts on dollars collected, the timeframe of the experiment would need to extend to the time the dollars are typically collected, because that is the experiment response measure.

5. Implementation plan: This section describes exactly how the pilot will be implemented, who is responsible for what, and what new training or written procedures will be required. This section is similar to a project plan, with the project as the pilot.

6. Tracking results: This is a critical component of an effective design plan. Exactly what is to be tracked, the source of the data, and the format of the tracking form need to be detailed and agreed upon. Operational definitions should be documented for all measures to be tracked.

7. Assumptions: This section lists any relevant assumptions involving the experiment, such as key personnel involved and required resources.

8. Risk assessment: A change in process typically carries some level of risk, especially an untried process. The types and level of risks are considered and documented in this section of the plan.

After the plan is completed and agreed upon, the next step is execution of the implementation plan, along with tracking of the data. Data from the experiment must be collected and initially evaluated early. This prevents wasted time running an experiment in which the data collection is not what was planned or needed. Any surprises need to be known and corrected early in the process. Once refinements to the initial plan are made and implemented, ongoing tracking of the experiment data can proceed.

I have included discussion of pilot experiments in the Improve Phase because that is when they are typically conducted. Pilot experiments are also conducted during the Control Phase, especially if the scope is wide and the time to track results is long. If the timeframe of the project is 4 to 6 months, and that much time is needed just to track the results, the pilot needs to be conducted during the Control Phase. If the pilot experiment results differ from expectations, the Six Sigma team may need to spend time following up and refining the project results.

9.6 COST/BENEFIT ANALYSIS

As we learned in the section on "The Goal" in Chapter 8 (Section 8.2), profit is the objective of any business. Six Sigma is a proven approach to increase profits, primarily through reducing inventory, work in process, non-value-added activities, and cycle time. To substantiate an improvement recommendation, the Six Sigma team must conduct a thorough and logical analysis of the expected costs and benefits of the change. Of course, at this point in the project, the team can only estimate total savings, based on the COPQ measurements discussed in Chapter 7. Tracking of actual results occurs in the Control Phase.

The cost/benefit estimates likely drive the Improve Phase choice of alternatives. It is common sense that a process change with a huge projected net benefit would be favored over a smaller benefit change. However, the team's decision making must account for risk that may or may not be implicit in the financial cost/benefit calculations.

The benefits of a proposed change may seem too difficult to measure. Although in transactional services it is more difficult to estimate benefits than in a manufacturing environment, it is certainly possible. Six Sigma is a business initiative, requiring that bottom line benefits result from most projects. As long as rational and logical assumptions can be made, estimating benefits is feasible, regardless of how intangible. In most service direct-customer-contact environments, such as a call center, changes made to increase customer satisfaction almost always results in decreased costs — the difficulty is quantifying the savings. In general, process changes that are good for the customer are good for the business, especially in a call center environment. Poor customer service results in higher expenses, primarily because of call backs.

9.7 IMPLEMENTATION PLAN

After the improvement recommendations have been agreed upon and the cost/benefit analyses completed, it is time to develop an implementation plan, detailing what steps are necessary to implement the changes, the timeframe of implementation, who is responsible, and any assumptions made regarding implementation. The project Champion and business line managers affected by the changes have likely been included in the project — if not in all phases, at least in the Improve Phase. The business line managers should take a lead role in developing the implementation plan, because they are responsible for making the changes happen.

The implementation plan is a project plan, very similar to the Six Sigma project plan described in Chapter 6. The difference, of course, is that rather than planning steps to carry out a Six Sigma project, the plan is for carrying out the tasks to implement the changes resulting from the project. The role of the Black Belt, the Green Belt if applicable, and the other Six Sigma team members in implementing the changes varies from company to company and from project to project. The most likely role for these individuals is maintaining the control plan, which is discussed in the next chapter.

9.8 SUMMARY

The Improve Phase of a Six Sigma project is where the project finally comes together. Depending on the scope of the project and the scope of the changes, this may be the most challenging, yet interesting phase. It is time for the left-brain thinkers who have performed superbly in the left-brain phases of Measure and Analyze, to switch their dominant thinking to the creative right side. A true mark of intelligence is to use both sides of the brain to perform complex, highly analytical tasks and then marry the acquired knowledge with creative and innovative solutions. The odds of success in this endeavor increase if a multidisciplined team conducts the project. However, most of the responsibility for ensuring a quantitative approach followed by innovative solutions rests with the Six Sigma Black Belt. Certainly, the team is there to support the task, but the Black Belt's role is to ensure success of the project. Some projects will achieve success through incremental improvements and others through creative and innovative ideas that result in dramatic improvements. The Improve Phase of the project determines which of these outcomes result from the project. I recommend taking the risky but innovative approach that strives for dramatic improvements, when possible.

9.9   CARD ONE CASE STUDY IMPROVE PHASE RESULTS

The results described below are from the Six Sigma Card One case study, which significantly increased the quality of service and decreased costs by an estimated $7.76 million per year:

1. Higher customer satisfaction ratings were achieved through lowering the average time it takes for customer service agent to answer a call. This was accomplished through implementing an improved staffing and scheduling system for the call center agents. How? The average time to answer a call, called average speed of answer, is directly correlated to staffing levels, because the more staff there is to answer incoming calls, the less time it takes for any call to be answered. Previously, there were two separate systems for forecasting call volume and required staffing and scheduling. A new automated staffing and scheduling system that allowed calls to be routed real time to either the Atlanta or Omaha call center was recommended. Staffing levels were decreased while service levels improved by 16% — from 76% of calls answered in less than 30 sec to 84% answered in less than 30 sec.

2. Through benchmarking, the Card One project team discovered that research has consistently proven that satisfaction levels in an inbound call center are strongly correlated with first contact resolution. The team developed an operational definition of "first contact resolution" as a call after which no additional contacts were required. Through analyzing the repeat call reasons, utilizing Pareto analysis and histograms, the team was able to determine that 38% of repeat calls were related to a case number generated when customers called in with a merchant billing dispute. When

their next billing statement arrived, there was a notice printed on the statement with the case number, but with little explanation of what it was about. The case number was printed on the statement even though the dispute may have been resolved in the interim. Providing a confirmation number at the time of the original inbound call and eliminating the notice on the billing statement significantly reduced the number of repeat calls. This is a good example of becoming more customer centric. By taking an "outside looking in" perspective of their business processes, the project team realized that "setting up a case" only had meaning for Card One employees. They addressed the issue by using industry-standard terms (e.g., confirmation number) rather than internal company jargon when communicating with their cardholders. The team estimated an 80% reduction in the 38% of repeat calls due to this type of inquiry.

3. The Six Sigma team dramatically reduced recruiting and training expenses by increasing the effectiveness of the recruiting, hiring, and training processes for new call center agents. Call center agent preselection tools for better hiring decisions and enhanced new hire training classes lowered expenses by reducing turnover and increasing agent productivity. At the same time, customer satisfaction increased because better selected and better trained agents were handling the calls. Turnover was estimated to decrease by 27% (from 42% to 33% per year), resulting in recruiting and training expense reductions of $48,000 per year. Further, the team estimated a 6% reduction in repeat calls and transfers because of more experienced agents handling calls.

The financial results of all the improve recommendations are summarized below:

Call volume was reduced by 11.6% through reducing the number of repeat calls and transfers to other departments.
- 9.3% of the reduced calls were through changing the process for setting up a billing dispute case.
- 2.3% of the reduction was through reducing turnover of agents, resulting in less repeat calls and transfers to other departments through a more experienced agent staff.

11.6% of 18.36 million calls per year = 2,129,760 calls eliminated per year @ $3.62 per call = $7,709,731 annualized savings plus reduced recruiting and training costs of $48,000 per year = $7,757,731 total savings.

Although most Six Sigma teams will not have the opportunity to achieve these kinds of dramatic results, the case study demonstrates that using the Six Sigma DMAIC approach, along with the analytical tools, has great potential to help any company become more profitable.

REFERENCES

1. Noguera, J. and Nielson,T., "Implementing Six Sigma at Interconnect Technology," ASQ FAQ Series on Six Sigma, 2000, from a presentation at the 46th Annual Quality Congress, 1992.
2. Hammer, M. and Champy, J., *Re-Engineering the Corporation: A Manifesto for Business Revolution*, Harper Business, New York, 1993.
3. "What Is a Survey?" http://www.amstat.org, American Statistical Society, 2001.

10 Control Phase

10.1 OVERVIEW

A Six Sigma team typically has a tollgate review after completing the Improve Phase and before the Control Phase, to ensure project stakeholders are on board with the improvements and ready to commit to the changes. The project sponsor, Champion, team members, and other stakeholders should attend this critical tollgate review. The project team, along with the Champion, Black Belt, or Green Belt, present the findings of the project and the final recommendations for improvement. It is only after final agreement and approval at this review does the implementation plan commence.

After the tollgate review and any final modifications to the project recommendations resulting from the meeting are made, the changes start to happen. The responsibility for the implementation rests largely with the Champion and business line managers whose areas the changes affect. The role of the project team takes on that of assisting in monitoring the success of the changes through active administration of the control plan. Whether the entire team is responsible for ongoing administration of the control plan, or it is limited to one or two individuals, the project Black Belt must maintain at least minimal involvement throughout the entire process. Closing the loop on a project is essential to the ongoing success of the company-wide Six Sigma effort, and the Black Belt is the most likely candidate to ensure this happens.

The objective of the Control Phase is to determine whether the expected improvements actually occur. This is accomplished through several tools and techniques, which are covered in this chapter:

1. Control plan
2. Process scorecard
3. Failure mode and effects analysis (FMEA)
4. Statistical process control charts
5. Final project report and documentation

The primary tool in this phase is the control plan. A process scorecard is a reporting tool that provides a quick summary of the key process measures. The FMEA helps the team to understand the risks associated with implementing a new process and proactively plans what to do in case of failures. SPC charts were covered in Chapter 7; a review of that material is provided in this chapter, because control charts are the primary tool used in ensuring the improvements are sustained.

10.2 CONTROL PLAN

A control plan is a reporting tool for tracking the performance measures and CTQs identified in the project for a period of time after the project, to measure expectations against actual results. In addition to the improvement measures, assumptions made in the Improve Phase regarding cost/benefit analyses are included for ongoing tracking; otherwise, it is difficult to know whether the actual project savings are realized.

The control plan is typically divided into two sections. The first is for ongoing tracking of the customer CTQs, which are usually the critical Y outcomes of the process. The second section is for the input and process measures, which are generally the business CTQs. Figure 10.1 provides an example of a control plan for the Card One case study.

10.3 PROCESS SCORECARD

A process scorecard provides a summary of the performance of the critical process measures, including the CTQ measures. Process scorecards are similar in concept to balanced scorecards, which is a reporting tool that provides key organizational performance measures. Robert Kaplan and David Norton proposed the concept of balanced scorecards in a 1996 book entitled, *The Balanced Scorecard: Translating Strategy into Action.*[1] The balanced scorecard measures four key areas of results: (1) financial, (2) customer knowledge, (3) internal business processes, and (4) learning and growth. It helps the company focus on key strategic measures required for organizational success.

The process scorecard is more tactical than the organizational balance scorecard. As indicated on the bottom left side of the control plan in Figure 10.1, scores are provided for each measure, depending on its performance relative to the target. This information is translated into a monthly scorecard that tracks not only the success of the Six Sigma project, but also the key performance measures of the department.

10.4 FAILURE MODE AND EFFECTS ANALYSIS

FMEA is used to prioritize potential defects and process failures based on their severity, expected frequency, and likelihood of detection. It is used in the Control Phase to understand the potential risks in implementing a new process and includes plans to respond to the risks if necessary. Not all new process implementations require an FMEA, although it is recommended that changes involving customer CTQs have some sort of FMEA in place to ensure customers are not negatively impacted by the changes.

The FMEA process begins by identifying all of the probable failure modes. This analysis is based on experience, review, and brainstorming, and actual data if possible. The next step is to assign a value on a 1–10 scale for the severity, probability of occurrence, and probability of detection for each of the potential failure modes. The three numbers for each failure mode are multiplied together to yield a risk priority number (RPN).

Card One Six Sigma Control Plan

Customer Critical to Quality (Outcome) Measures	Target	Baseline	Month1	Month2	Month3	Month4	Month5	Month6	Month7	Month8	Avg
1. First contact Resolution %											
2. Quality score %											
3. ASA in seconds											
4. % Calls within 30 secs											
Input & Process Measures											
1. Inbound calls											
2. IVR complete %											
3. Abandoned %											
4. AHT in seconds											
5. Agent turnover %											

Score Key:	Relative to Target
A+	= >110%
A	meets
A-	= >95%
B+	= >90%
B	= >85%
B-	= >80%
C+	= >75%
C	= >70%
C-	= >65%
D	< 65%

Measurement Definitions

Customer CTQ Measures	Input and Process Measures
1. First contact resolution = % of calls with no follow-up contact	1. Inbound calls: total calls
2. Quality score: aggregate quality score as measured by monitoring	2. IVR completes %: Inbound calls completed by IVR over total inbound calls
3. ASA in seconds: avg speed of answer	3. Abandoned %: # of abandoned CRD calls divided by total CRD calls
4. % Calls answered within 30 secs	4. Average handling time (AHT): average talk time + wrap time
	5. Agent turnover %: average # of agents divided by # of agents terminated (voluntary & involuntary)

Card One Customer Service Department

FIGURE 10.1 Example of control plan for the Card One case study.

The RPN becomes a priority value to rank the failure modes, with the highest number demanding the most urgent improvement activity. Figure 10.2 illustrates an FMEA template, Figure 10.3 provides criteria for the severity rating, Figure 10.4 provides criteria for ranking the probability of the failure occurring, and Figure 10.5 lists the detection failures.

The steps involved in performing an FMEA are listed below:

1. Create worksheet and develop operational definitions for the potential failures.
2. Interview appropriate employees or conduct group brainstorming sessions to determine what the failures are and their likelihood of occurrence.
3. Rate the potential effect of the failures on a scale of 1 to 10.
4. Rate the severity of the potential effects.
5. Rate the probability of detecting the failures.
6. Calculate the RPN (multiply results from #3, 4, and 5).
7. Develop action plans for each failure mode, giving prioritization to the highest RPN.

Defining failure may seem trivial, but it is an essential step in the analysis. It is important to operationally define what a failure is, because if we were to ask 50 people to define failure, we would probably get 50 different definitions (operational definitions at work). Some common definitions of failure are:

- Any loss that interrupts the continuity of the process
- A significant loss of company assets, such as employees or market share
- Unavailability of equipment
- An out-of-control situation
- Not meeting target expectations

10.5 SPC CHARTS

As discussed in Chapter 7, control charts have two basic purposes. The first is to *analyze past data* to determine past and current performance of a process. The second is to *measure control of the process against standards*. In the Measure Phase, control charts are used primarily for the first purpose of determining past and current performance of a process.

The second purpose of control charts, ongoing measurement against standards, is applicable in the Control Phase. Control charts at this point determine whether the intended result of a process change (the standard) actually happens, particularly changes intended to have a positive effect on the process outcome measures.

I will not repeat the full treatment of control charts in this chapter. The reader is directed to Chapter 7 for a full discussion of the background, applications, and benefits of control charts. However, a review of the basics of constructing control charts, along with a description of the different types of control charts, are included here.

Process Step or Activity	Potential Failure Mode	Potential Effect(s) of Failure	S e v	Potential Cause(s) of Failure	O c c u r	Current Controls	D e t e c t	R P N	Recommended Actions	Responsibility & Target Completion Date	Actions Taken	S e v	O c c u r	D e t e c	R P N
												0	0	0	0
												0	0	0	0
												0	0	0	0
												0	0	0	0
												0	0	0	0
												0	0	0	0
												0	0	0	0
												0	0	0	0
												0	0	0	0
												0	0	0	0
												0	0	0	0
												0	0	0	0
												0	0	0	0
												0	0	0	0

FMEA Type (Design or Process):

Responsibility:

Core Team:

Project Name/Description:

Prepared By:

Date (Orig.):

Date (Rev.):

Date (Key):

FIGURE 10.2 FMEA worksheet.

Severity Evaluation Criteria Example for Process FMEA		
Effect	Criteria: Severity of Effect	Ranking
Hazardous — without warning	If involves noncompliance with government regulation. Failure will occur without warning.	10
Hazardous — with warning	If involves noncompliance with government regulation. Failure will occur with warning.	9
Very high	100% of output is defective. Customer dissatisfied.	8
High	A portion of output is defective. Customer dissatisfied.	7
Moderate	A portion of output is defective. Customer experiences discomfort.	6
Low	100% of product will have to be reworked. Customer experiences some dissatisfaction.	5
Very low	A portion of output may have to be reworked. Defect noticed by some, but not all customers.	4
Minor	A portion of output may have to be reworked. Defect noticed by average customers.	3
Very minor	A portion of output may have to be reworked. Defect noticed by discriminating customers.	2
None	No effect.	1

FIGURE 10.3 Severity evaluation criteria worksheet.

Occurrence Evaluation Criteria Example for Process FMEA			
Probability of Failure	Possible Failure Rates	Cpk	Ranking
Very High - Failure is almost inevitable	>=1 in 2	<0.33	10
	1 in 3	>=0.33	9
High - Generally associated w/processes similar to previous processes that have often failed.	1 in 8	>=0.51	8
Moderate - Generally associated w/processes similar to previous processes which have experienced occasional failures, but not in major proportions.	1 in 80	>=0.83	6
	1 in 400	>=1.00	5
	1 in 2000	>=1.17	4
Low - Isolated failures associated with similar processes	1 in 15,000	>=1.33	3
Very Low - Only isolated failures associated with almost identical processes.	1 in 150,000	>=1.50	2
Remote - Failure is unlikely. No failures ever associated with almost identical processes	<=1 in 1,500,000	>=1.67	1

FIGURE 10.4 Occurrence evaluation criteria worksheet.

The basic construction of a control chart is a center line (CL) and the upper and lower control limits (UCL and LCL). The center line is the average value of many data sets tracked over time, and the UCL and LCL are set to three standard deviations from the center line. Only about 0.027% of observations will go beyond the upper and lower control limits when the process is in a state of statistical control.

The first step in utilizing a control chart is to determine the measure of interest and whether it is continuous or discrete data. The data are then charted to estimate

Detection Evaluation Criteria Example for Process FMEA		
Detection	Criteria - Likelihood that the Existence of a Defect will be Detected by Process Controls before Next of Subsequent Process	Ranking
Almost impossible	No known controls available to detect failure mode	10
Very remote	Very remote likelihood that current controls will detect failure mode	9
Remote	Remote likelihood that current controls will detect failure mode	8
Very low	Very low likelihood that current controls will detect failure mode	7
Low	Low likelihood that current controls will detect failure mode	6
Moderate	Moderate likelihood that current controls will detect failure mode	5
Moderately high	Moderately high likelihood that current controls will detect failure mode	4
High	High likelihood that current controls will detect failure mode	3
Very high	Very high likelihood that current controls will detect failure mode	2
Almost certain	Current controls are almost certain to detect the failure mode. Reliable detection controls are known with similar process.	1

FIGURE 10.5 Detection evaluation criteria worksheet.

the common cause variation. This is done by sampling small subgroups of 2 to 10 consecutive observations from an ongoing process and treating each subgroup as separate data sets. The subgroup averages and range are plotted over the time period and if an out of control signal is detected, such as one or more points falling outside the three sigma control limits, further action is taken to determine the cause of the out-of-control signal. There are many special software packages on the market exclusively for SPC applications. Minitab also has excellent SPC capabilities and is very easy to use. The key is to pick the appropriate chart for the situation.

Seven common control charts are typically used in the Control Phase. Their use is dictated by the type of data (continuous or discrete), the sample size, and the type of defect measures. For continuous variables:

1. X-bar and R charts: Used for continuous outputs such as time and weight. The most common type of continuous output measure in transactional services is duration of time, such as time to process an invoice or cycle time of a purchase order. The X-bar represents the subgroup averages and R is the range within each subgroup. The X-bar chart tracks shifts in the process mean and the R chart tracks the amount of variability in the process.
2. X-bar and S charts: Similar to the X-bar and R charts except that the standard deviation (instead of the range) is used as the dispersion measure. When subgroup sizes are greater than 10, an S chart should be considered.
3. X and mR charts: When rational subgroups are not feasible, such as very long intervals between observations, X and mR charts are used. These charts are also called Individuals and Moving Range charts. Individual observations (subgroup size = 1) are the X's plotted over time and the moving range of the successive data points is used as the range measure to calculate control limits of X.

For discrete variables:

1. nP charts: These charts are used for tracking the proportion of defective units when the sample size is constant. The output is classified as either defective or not, similar to a pass–fail situation.
2. P charts: This is the same as the nP chart, except the sample size varies.
3. C charts: C charts are used when there is more than one defect per unit and it is calculated as the number of defects produced when each unit has the same opportunities for defects (e.g., a constant sample size).
4. U charts: Similar to C charts, U charts are used to track defects per unit, but the unit is of varying size. For example, if the number of errors on an invoice is measured, and the fields of data on the invoice are constant from invoice to invoice, then a C chart is appropriate. A U chart is appropriate for measuring the number of errors in outgoing correspondence that is of variable size and contains data that differ among letters.

The signals for an out-of-control situation are:

1. One or more plots outside the upper or lower control limits
2. At least two successive points fall in the area beyond 2 standard deviations from the center line in either direction
3. At least four successive points fall outside the 1 standard deviation line
4. At least eight consecutive values fall on either side of the center line

Control charts should be developed and maintained on an ongoing basis for all the key measurements in the control plan. When the measurement results are entered into the control plan worksheet, the SPC charts should also be updated.

10.6 FINAL PROJECT REPORT AND DOCUMENTATION

Many companies have project databases that serve as repositories for Six Sigma project results. If this is the case, the final project report and all of the associated documentation will need to follow the repository's specific requirements. Other companies may have standardized templates for reporting on Six Sigma projects, along with standard guidelines for final project documentation. In the absence of a repository or formalized project templates, the Six Sigma team needs to develop its own standard project documentation.

Good project documentation is an essential step in closing the loop between the base knowledge level the team started out with and the knowledge gained through conducting the project. Documentation is also important for the record — to make sure the improvement rationale is available for future employees and improvement teams, as well as to share the knowledge from the project with interested parties.

There are several components to the final Six Sigma project report. The report should be divided into the DMAIC Phases, with appropriate documentation. Below are suggested components; however, each project is different, thus the project report contents may vary from project to project:

1. Introduction — Provides the reader with a background of the project and the project selection process.
2. Six Sigma project plan.
3. Define Phase — Include the project charter, VOC, and the S-I-P-O-C.
4. Measure Phase — Include list of potential critical X's and Y's, baseline performance measures, and summaries of key results from application of measure tools, such as Pareto charts, histograms, and control charts.
5. Analyze Phase — Include results of all statistical tests and the conclusions drawn from them. Include root cause analysis documentation, such as C&E diagrams.
6. Improve Phase — Include final proposals for improvement and the rationale or data analysis supporting the recommendations, such as a pilot plan and results. Include a cost/benefit analysis of the improvement recommendations.
7. Control Phase — Include a copy of the control plan, FMEA, and any other relevant material, such as process scorecards.
8. Implementation plan.
9. New process documentation.
10. Actual control plan, updated regularly with process performance data.

REFERENCE

1. Kaplan, R.S. and Norton, D.P., *The Balanced Scorecard: Translating Strategy into Action*, Harvard Business School Press, Boston, MA, 1996.

11 Design for Six Sigma

11.1 OVERVIEW

There are generally two situations in which a Design for Six Sigma (DFSS) approach is preferred over the DMAIC approach:

1. If there is not an existing process in place to improve upon, or if the existing process is so broken (i.e., less than one sigma) that it is not worth the time and effort to improve it
2. If the process has been improved enough to reach a high sigma level (five sigma or greater) and there is no way to increase the sigma level under the current process parameters

DFSS is a structured, disciplined approach to build new products and processes that focuses on customer CTQs. Like DMAIC, DFSS uses a phased project management approach. DFSS is used for both product and process design; although the approach to both is similar, this chapter focuses on process design. The four DFSS phases are described below:

1. *Define Phase*. Similar to the DMAIC, this phase sets the foundation for the project, along with goals and project plans. The deliverables of this phase include:
   - Project charter outlining business case, design goals, and expected benefits.
   - Project plan including tasks, responsibilities, and due dates.
   - Customer CTQ requirements. The tool most often used to capture customer CTQs and translate them to specific requirements is quality function deployment (QFD).
   - Business CTQ requirements. These include marketing and competitive requirements, resource constraints, and the return on investment requirements of the business
2. *Conceptual Design Phase*. Alternative designs are developed, flow charted, evaluated, and tested in this phase. Creative problem-solving techniques, such as Theory of Inventive Problem Solving (TRIZ), are utilized. FMEA and cost/benefit analysis are used as filtering mechanisms to select an optimal design.

3. *Optimize Design Phase.* The selected design is optimized in this phase through testing and piloting against CTQs. Another FMEA is developed to understand the design weaknesses. Another design iteration is performed to further optimize, followed by stress testing to test for robustness and yet another design iteration to further optimize. Documentation and training materials are developed on the new process. A small-scale pilot experiment is conducted if feasible.

4. *Implement and Control Phase.* The new process is implemented and tracked against a control plan to ensure control and adherence to standards. Any necessary training and new support systems are implemented. Close monitoring of the new process is conducted during the early phases, and refinements are made as necessary. Process outputs are tracked and measured against customer and business CTQs.

DFSS differs from traditional developmental approaches in the same way that DMAIC differs from traditional quality and process improvement efforts. The primary differences between DFSS and past design efforts are described below:

1. The percentage of time devoted to front-end planning is higher as a percentage of total development and implementation time. The idea is to solve the problem at the root or upstream in the process rather than later downstream. This is applicable to both the DFSS process as well as the operational process that is being designed. DFSS quality early in the design process prevents many costs from being incurred downstream.

2. The focus is on gathering and verifying customer CTQs before design actually starts. Checkpoints along the design process ensure that CTQs are incorporated at every stage. Operationally defined requirements, driven by customer CTQs, ensure that there will be little or no extensive rework of the design process further downstream.

3. Consideration of alternative design concepts and logical selection of the optimal design minimize the chance that design changes will be required after "live" implementation.

11.2 DFSS TOOLS

DFSS uses a variety of tools, including process mapping, cost/benefit analysis, FMEA, QFD, and TRIZ. Chapter 7 covered process mapping, Chapter 9 covered cost/benefit analysis, and FMEA was discussed in Chapter 10. This chapter, therefore, will focus on QFD and TRIZ.

11.2.1 QUALITY FUNCTION DEPLOYMENT

QFD is a tool to capture and translate customer requirements into technical design requirements. It is also called "house of quality" because the end product looks similar to a house with a steeple roof. The technique was developed by Yoji Akao, who first applied it at the Mitsubishi Heavy Industries Kobe Shipyard in 1972. Toyota

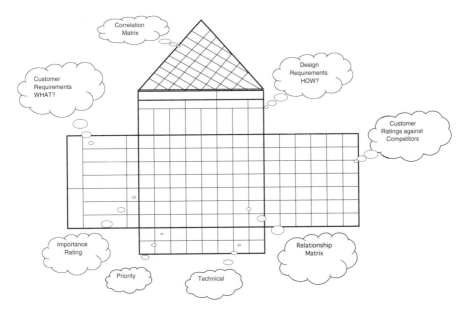

FIGURE 11.1 QFD template.

adopted the tool for its U.S. factories in the late 1970s, eventually reducing product development costs by 61% and cycle time by 33%. Akao defined QFD as "a method for developing a design quality aimed at satisfying the consumer and then translating the consumer's demands into design targets and major quality assurance points to be used throughout the production phase."[1]

QFD uses actual statements expressed by customers to capture VOC, then translates those statements into customer requirements, which are used to develop technical design requirements. Figure 11.1 provides a template for a house of quality.

The data required to complete a QFD are typically gathered through interviews; however, surveys and benchmarking are also good sources of QFD data. They can be broken down as follows:

1. Customer requirements (what?): This section represents the VOC, listing their CTQ requirements. Regulatory and business CTQs can also be listed here.
2. Importance rating: The importance of each CTQ is ranked from 1 (low) to 5 (high). The customer assigns these ratings.
3. Design requirements (how?): These are the technical requirements associated with meeting the various CTQs.
4. Relationship matrix: Symbols in the relationship matrix represent the degree to which the design requirements are necessary to fulfill the CTQs. Three different symbols are used, each corresponding to a number: 9 equals a strong relationship, 3 equals some relationship, and 1 equals little or no relationship.

5. Customer ratings: This section contains ratings obtained from the customers on their perceived quality of the competition. If this is a DMAIC project, the current process can be rated along with the competition; in DFSS projects, there is usually no current process to rate.
6. Priority: This value is calculated by multiplying the importance rating by the relationship indicator. It measures the priority of a particular technical design requirement relative to other design requirements.
7. Technical difficulties: This is a rating assigned by the project team on the difficulty of building the technical requirement; this is an indication of the expected effort that will be required in meeting the requirement.
8. Correlation matrix: This is the roof of the house. It is used to identify where technical requirements support or impede each other.

Figure 11.2 provides an example of a QFD chart for the Card One case study. Card One plans to add a bill payment option to the Card One customer service internet site. QFD was used to understand VOC in designing and building the new site pages. Let us review each part of the QFD in Figure 11.2 as it relates to the case study:

1. Customer requirements: The two primary CTQs, listed on the left side of Figure 11.2, are functionality and ease of use of the bill payment option. Within each category, there are particular requirements with an importance rating assigned to each (1 = lowest, 5 = highest).
2. Importance rating: The ability to pay bills with a debit card tied to a checking account or with a check is the #1 CTQ for the new pages. Providing a confirmation number and access to the bill payment screen from any other page in the site are also rated highly.
3. Technical design requirements: Translating the CTQs into technical design specifications provides us with the how's of the QFD. Distinctions are made between functional and design technical requirements.
4. Relationship matrix: A relationship correlation indicator is provided between the CTQ and the technical requirement, from 9 (strongly related) to 1 (little or no relationship).
5. Customer ratings: The DFSS team has chosen American Express and First City VISA as benchmarks. Customer input is obtained to rate the customer's perception of how well these competitor companies' bill payment options perform against the CTQs.
6. Priority: A priority rating for each technical requirement is calculated by multiplying the importance rating by the relationship indicator weight, such as multiplying the importance rating of 3 for "can schedule pay date" by 9 for "calendar storage and retrieval" to arrive at a priority of 27. This value provides the DFSS team with an indication of the importance and, thus, the priority of a particular technical requirement to the overall design of the new bill payment option.
7. Technical difficulty: An assessment of the technical difficulty of each technical requirement is arrived at through team consensus and subject matter expert advice.

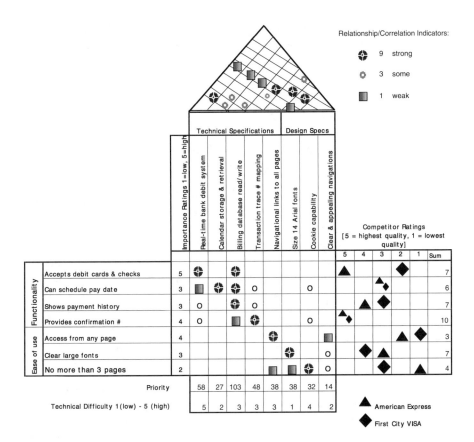

FIGURE 11.2 QFD for designing the Card One customer service internet bill payment option.

8. Correlation matrix: We see strong correlations between the functional requirements and also between the site design requirements. Only minor correlation is found between the functional requirements and site design requirements.

The QFD results are used to develop the DFSS project plan for scheduling priority and resources. If resource constraints dictate selection of certain CTQs, the QFD ensures the VOC will be heard loudly and clearly in choosing among alternatives. The priority rating along with the expected technical difficulty in meeting the technical requirements are used to prioritize and schedule the project plan tasks. The need for direct and real-time updates between Card One's primary billing database and the internet server data is the number one priority. The requirement for a system to communicate in real time with various banking networks and the ability to tag each transaction through a trace number are the second and third priorities, respectively.

QFD has not gained widespread use in the service industries' quality initiatives compared to other Japanese imported tools. The reason most often cited is the differences in business cultures between American management practices and the Japanese. I am not convinced of this and suggest the more likely reason is a need for education and training, similar to DOE. As the Six Sigma approach of using a disciplined and fact-based discovery process to make better decisions is further ingrained in U.S. business practices, we will likely see broader use of QFD, especially for DFSS projects.

11.2.2 THE THEORY OF INVENTIVE PROBLEM SOLVING (TRIZ)

TRIZ (pronounced "treez") is a structured problem-solving approach aimed at tapping into the creative right side of our brains.[2] It has great appeal among Six Sigma practitioners because of its data-based approach to innovation and creative problem solving. The method originated through the work of Genrich Altshuller, a young travel officer in the Soviet Union after World War II. After the war, 20 million Russians were dead and 50% of their industry had vanished. Altshuller was tasked with applying the discoveries and innovations from the war effort to help with the massive rebuilding effort.[2] He had the insight to realize there was no time to tinker around with ineffective or ordinary inventions, so he attempted to methodologically distinguish between ordinary and extraordinary discoveries. Through his research in innovative patents, Altshuller made three important discoveries: (1) problems and solutions are repeated across industries and sciences; (2) patterns of technical evolutions are repeated across industries and sciences; and (3) innovations used scientific effects outside the field where they were developed.

The theory behind TRIZ is the fact that at the heart of a problem is a contradiction. In order to solve a problem, trade-offs are involved, and by definition, trade-offs result in suboptimal solutions. TRIZ is about getting rid of the problem. Just as DFSS involves significant structured planning efforts early in the process, TRIZ applies significant up-front effort to redefine the problem in view of proven patterns of problems and solutions. Rather than solutions reached through spontaneous creativity, TRIZ reaches solutions through focused, structured creativity based on a number of principles. The TRIZ problem-solving path follows (1) the specific problem at hand to (2) a TRIZ general problem to (3) a TRIZ general solution to (4) the specific problem solution. Analytical methods and various tools are used to move from the specific to the TRIZ general problems and solutions. The reader is referred to the TRIZ journal at www.triz-journal.com to learn more about TRIZ methods.

TRIZ distinguishes between two types of contradictions: technical and physical. *Technical contradictions* address the trade-off problem typically seen in new product and process design projects; as one aspect of the design gets better, another gets worse. A good example is in designing a sales forecasting process, the more data used in the forecast, the more accurate it is. However, the trade-off is that more data are more expensive and time consuming. Another example is in order to provide excellent service to inbound callers at Card One, more staffing and training are required, which increases costs.

Physical contradictions involve problems where the solutions are in direct conflict with each other. An everyday example is the fact that food with high fat content tastes good but is high in calories, leading to excess weight gain. A business example is that inbound call center agents should provide thorough, high-quality answers to customers on the first contact, but should minimize the time spent on each call. The boundaries between the two categories of contradictions are not always clear; in general, the technical contradictions involve more concrete solutions whereas the physical contradictions involve more comprehensive solutions.

One of the most often used and popular tools, especially for the TRIZ novice, is the 40 principles of inventive problem solving.[2] The 40 principles are used to identify and resolve technical contradictions and are based on the repeating patterns of solutions and technical evolutions originating from TRIZ. Appendix C provides an abbreviated list of the 40 principles.

Physical contradictions are addressed through four "principles of separation":

1. Separate in time
2. Separate in space
3. Phase transition, such as moving from a solid to a liquid to a gas
4. Moving to the supersystem (macro) or to the subsystem (micro)

The TRIZ journal uses the example of an airbag to illustrate using the principles of separation to solve the contradiction that airbags should deploy at both a high threshold and a low threshold.[2] Applying the four principles leads to solving the root cause of the problem that airbag deployment causes injuries, rather than merely addressing the symptom of decreasing the speed of deployment. The principle of separation in space leads us at first down the road of solutions, such as using sensors to detect the size of the passenger and adjust accordingly, or to separate small children from the airbag by placing in the back seat. However, these do not get rid of the problem that airbags cause injuries. The principle of space leads us to examine the space where the problem occurs. Applying the TRIZ law of technology evolution, which states that systems become more segmented over time because smaller segments are easier to control, leads to the solution of using micro airbags. Micro airbags do not cause injuries but are sensor controlled to adjust to the size of the person. The root cause of injuries is eliminated while the person is protected by the airbag deployment. TRIZ has been successfully used in research and design for a wide variety of applications. Examples include reducing warranty costs at Ford Motor Company and information technology product development at DelCor Interactive International. This section has merely touched on a few of the TRIZ methods for creative problem solving and design. The interested reader is directed to www.triz-journal.com for a wide variety of topics on TRIZ.

REFERENCES

1. University of Sheffield QFD web site, 2001.
2. www.triz-journal.com, 2001.

12 Introduction to Lean Servicing™

12.1 LEAN PRODUCTION OVERVIEW

Lean Production, also known as Lean Manufacturing, is a business philosophy that was originally developed at the Toyota Motor Company. Thus, the Toyota Production System (TPS) is synonymous with Lean Manufacturing. The objective is to eliminate all forms of waste in the production process. Taiichi Ohno, the "father of lean," identified seven forms of waste (*muda* in Japanese):[1]

1. Overproduction — using production equipment and machines faster than necessary without regard to whether the output of the machines is required downstream.
2. Waiting for machines or operators — operators may wait for available machines or machines may wait until an operator is available.
3. Transportation waste — unnecessary movement of parts or people around the production facility.
4. Process waste resulting from inefficient, poorly designed processes — duplication of effort, inspections, and nonvalued activities.
5. Excessive inventory — unnecessary work in process and finished goods beyond what is needed on a normal basis to keep the business running. The "evils of inventory" result in wasted floor space, capital, administrative costs, and storage costs.
6. Wasted motions through operators leaving workstations to fetch required supplies or through continuous reaching, searching, or carrying goods.
7. Waste of rework through producing defects.

When we eliminate all waste, the order production cycle time (time from receipt of order to receipt of payment) is compressed. The result is short cycle and delivery times, higher quality, and lower costs.

12.2 LEAN HISTORY[2]

The roots of lean go back to 1903 when Henry Ford produced his first automobile — the Model A.[3] It started with the need for closer tolerances of thousands of disparate parts, a must for success in mass production.

The craft production system that was previously used to build automobiles relied on skilled craftsmen who built each part by hand and skilled fitters who assembled the parts into the final product. The fitter's job was one of filing down individual parts so that they all fit together to produce a car. Every part and every car was built one at a time by hand. No two cars were alike; in fact, the cars differed considerably because each was a prototype. Of the estimated 1000 cars produced per year in the late 19th century, only about 50 were even of the same design. Because each car was hand built, usually to the customer's specifications, production costs were prohibitively high for all but the very wealthy.[4]

Ford's brilliant idea of *interchangeability of parts* was the genesis of mass production. To achieve this interchangeability, a more precise measurement system was needed. Each part had to be produced according to a formally documented measurement system. Because variability of each part was substantially reduced, the skilled assembly fitters could be replaced with less-skilled, lower-paid workers who worked much faster.[5] The *simple design* of the car resulted in reduced number of parts resulting in cars that were less complex than previous models.

Ford's first attempt at mass production involved a shop environment, where each car was built from beginning to end in a stationary environment. Two or three assemblers per car were responsible for building large sections, such as the chassis or the body. The assemblers would get their needed parts, work on one car for 6 to 9 h, and then repeat the process on the next car down. One of the first improvements to this process was to deliver the parts to the cars rather than having the assembler retrieve them, resulting in each assembler remaining in one place most of the day.[5]

The next major improvement was what we know as *specialization of labor*. The work of the assemblers was broken down into small, short-cycle tasks, such as bolting the wheels onto the chassis. Each assembler would go from car to car performing his specialized task, which took anywhere from 5 to 10 min. Productivity increased dramatically from shorter learning cycles and subsequent perfection of the specialized task. Because the work required little or no training, unskilled immigrants were hired at lower wages than the semi-skilled assemblers, resulting in lower costs to produce each car. However, the uneven task times created bottle-necks, and fatigue was an issue for the assemblers, who were walking from car to car, covering a significant distance in a day.[6]

This brings us to 1914 and the Highland Park Model T plant. Ford, on observing cow carcasses moving on a conveyor in a slaughterhouse, reasoned that if it worked for cows, it would work for automobiles.[7] He directed the new plant be built with a moving assembly line; now, instead of the assemblers moving to each stationary car, the cars were brought to the stationary workers. Almost a 100% increase in productivity was achieved with this one change.

The issues involved in maintaining close tolerances of thousands of disparate parts from many different suppliers led Ford to bring all suppliers in house for complete vertical integration. Vertical supply chain integration ensured availability of parts with relatively short lead times. However, other issues surfaced. The relatively long cycle time of machine set-up for making parts conflicted with the short production cycle of mass production. Ford solved this conflict by borrowing from his ideas of specialization of labor; in this case, it was specialization of machinery, specifically die machinery.[8] The huge machines dedicated to making a limited number of parts, arranged in sequential fashion, were the pinnacle of Ford's mass production system.

The hallmarks of mass production — interchangeability of parts, specialization of labor, and simple design — were taken to the extreme at Highland Park, resulting in a car that for the first time was affordable for many Americans — the Model T. Ford boasted that consumers could have the car in any color they liked, as long as it was black. Ford called the Highland Park assembly method continuous-flow manufacturing, a key concept in Lean Production. Indeed, Ford's early mass production system shares many commonalities with Lean Production. Other elements of mass production, however, are in stark contrast to lean.

When Taiichi Ohno was a young engineer with the new Toyota Motor Company in 1949, he visited Ford's River Rouge plant outside Detroit. He liked the idea of the continuous-flow manufacturing; however, other aspects of mass production appeared wasteful. The use of heavy and expensive press and die machines required heavy capital investment, which Toyota did not have. Die changes at the River Rouge plant required at least 1 day and were performed by die change specialists employed specifically for that purpose. This resulted in a large number of parts made between die changes, and thus changes were typically scheduled only four to five times a year. Ohno realized this system would not work for Toyota, not only because of the large capital investment required, but also because their production was only a few thousand vehicles per year, compared to millions produced in Detroit.[9,10]

After extensive experimentation, Ohno engineered a method to dramatically decrease the time and labor required to change dies. Whereas the typical U.S. auto plant scheduled die changes every 3 months, Toyota began scheduling changes every 2 to 3 h. The impact of this change was dramatic. Short production runs became the norm for Toyota, resulting in fewer inventories of finished goods on hand. And because the production cycle to produce one completed vehicle was shortened, quality issues were discovered sooner rather than after millions of parts had been produced, typical of U.S. factories. Earlier detection of defective parts translated to less rework, which drove down production costs while increasing the quality of the final product. This contradicted the theory of economies of scale, which was the basis for U.S. manufacturers' production systems.[11]

Through the years, Toyota has refined their Lean Production methods to become one of the highest quality producers of value-priced vehicles. We will explore these lean techniques and tools in the rest of this chapter.

12.3 LEAN SERVICING CASE STUDY INTRODUCTION

The Lean Servicing case study is the new credit card approval process for Card One. The new card application processing (AP) department is in Dallas and is housed in the same facility as merchant operations. Cards in force, voluntary and involuntary attrition, and card application volumes are shown in Figure 12.1.

The Dallas AP department is projected to receive approximately 9.4 million card applications in 2002 from a variety of sources, as shown in Figure 12.2. The cycle time from receipt of the signed applications to approval has steadily increased over

Card One Volume Statistics, 1998–2002

Year	Cards in Force	% Increase	% Attrition	# Cards Attrited	# Cards Retained	# Cards Added	% Acceptance	# Applications	% Increase in Apps
1998	6,557,760								
1999	7,286,400	10.0%	26.0%	1,705,018	5,581,382	1,705,018	30.0%	5,683,392	N/A
2000	8,096,000	10.0%	28.0%	2,266,880	5,829,120	2,266,880	33.0%	6,869,333	17.3%
2001	9,200,000	12.0%	26.0%	2,392,000	6,808,000	2,392,000	31.0%	7,716,129	11.0%
2002	10,396,000	13.0%	28.0%	2,910,880	7,485,120	2,910,880	31.0%	9,389,935	17.8%
Averages	9,230,667	11.7%	27.3%	2,523,253	6,707,413	2,523,253	31.7%	7,991,799	15.4%

FIGURE 12.1 Card One cards in force, card attrition, and card application volume.

Card One Application Statistics, 2002

Source of Application	Number of Apps	% of Total	Avg Cycle Time (Days)
Affinity card pre-approved applications	3,568,175	38%	11
Card One pre-approved applications	2,253,585	24%	9
Take one applications on display at merchant sites	2,065,786	22%	16
Internet applications	845,094	9%	13
Call-in applications in response to broadcast ads	657,295	7%	13
Total	9,389,935		11.94

FIGURE 12.2 Sources of applications for Card One.

the past 3 years, causing concern among Card One management and their Affinity card partners. A longer cycle time results in potential lost revenue, increased calls to Customer Service to check on the status, and image issues relative to Card One's competitors. The average weighted cycle time among all application sources is approximately 12 days, up from 10 days 3 years ago.

The AP department occupies the first two floors of a four-story leased facility; Card One merchant operations is housed in the upper two floors. The newly appointed Director of Applications Processing, Allen Shannon, is a trained Six Sigma Black Belt and has spent the last 2 years in the Atlanta Customer Service operation. After a few weeks on the job, he realizes that his new department is a factory in the true sense of the word — the manufactured product is a decision on whether to approve or reject the application. The factory's assembly line moves information from one workstation to another, hopefully adding value at each step in the process, until a decision is made. Allen remembers a new concept called Lean Production that was presented at a recent quality conference, recalling that Six Sigma and Lean Production are complementary approaches to achieve dramatic improvements in quality, cycle time, and costs.

Allen discusses his ideas with his manager, Harry Roskind, who is intrigued with the concept. They decide that Allen should attend an upcoming 2-week seminar on Lean Production that is specifically geared toward a transactional environment. The seminar is sponsored by Lean Servicing™, Inc.

After returning from his first week of training, Allen again meets with Harry to discuss a game plan. They decide that Allen will lead the department-wide effort but will need assistance from his staff. Anna Fredrick, a particularly bright and enthusiastic project analyst, is chosen to assist Allen in managing the project. Arrangements are made for her to attend the 2-week Green Belt training as well as accompany Allen on his second week of Lean Servicing training. They decide that outside consultants will be used only if necessary — Allen hopes the department can build the necessary talent internally, so that changes made through Lean Servicing are integrated into the work of the department rather than viewed as a one-time project.

After the second week of training, Allen again meets with Harry to lay out specific plans on how the transformation to a lean servicer will be accomplished, including resources and timing. Allen's goal to decrease the average card approval cycle time from 12 days to 3 days is met with disbelief. Harry wonders how Allen can achieve a fourfold improvement without bringing in a whole new system and hiring additional associates, which would add substantial costs to the process. Allen explains that not only does he intend to achieve this with no increase in headcount or costs, he is sure headcount and costs will *decrease* along with the cycle time. Although skeptical, Harry has always provided his managers with wide latitude and support — until they prove him wrong. He hopes that Allen proves to be right.

They decide to have a Lean Servicing company trainer provide a 1-day overview of lean concepts to the entire department of 302 associates. They make arrangements for a local hotel to host the seminar, which will be presented over 2 days: half the staff attending the first day and the remaining staff attending on day 2. Harry flies in from Atlanta and assists Allen in facilitating the seminar. The off-site location and the active participation of Card One management demonstrate to associates the significance of the effort. The seminar leader explains the importance of everyone's participation, because it is likely their jobs will change in a lean environment. Harry and Allen announce there will be no layoffs as a result of the changes and any headcount reductions will be through normal attrition. They explain that additional training will be required for everyone in the center in the upcoming months, because a principle of Lean Servicing is to staff with highly trained individuals who add direct value to the product. A new part of their job responsibility will be improving their jobs on a continuing basis through *kaizen* and Total Productive Maintenance (TPM) activities. The effort to transform the AP department into a lean servicer provider is informally dubbed "Lean, Mean, and Green."

The next section describes what Allen learned at the Lean Servicing training, and the subsequent section discusses how the concepts are applied to transform the AP department to Lean, Mean, and Green.

12.4 LEAN SERVICING CONCEPTS

The overall principle of lean is elimination of all kinds of waste. If this is accomplished, the only activities in producing a product or service are value-added activities. This is what Taiichi Ohno realized back in 1949. After many trips to U.S. auto manufacturing plants, Ohno was struck by the number of workers in the U.S. plants who were not directly involved in the making of the car. These included the plant foremen, industrial engineers, unskilled laborers, die change specialists, and so on. Through years of experimentation, Ohno designed the Toyota factory to eliminate as many non-value-added workers as possible. This required a cultural change from the mass production philosophy of extreme specialization of labor, where training for a new task was measured in minutes and communication among workers was a liability. In Ohno's new world, the assemblers took on the duties of all the nondirect workers as well. This meant continuous training of the workers, as well as agreements between the workers and management involving continuous improvement — the workers had a duty to continuously improve the process and management had a duty to provide ongoing training and to listen to and act on the workers' suggestions. The foreman position was replaced with a team leader, which was rotated among the team members; these work teams, called Quality Circles, worked together to continuously improve their work processes.[9]

Taiichi Ohno's dilemma in 1949 is a classic example of turning a problem into an opportunity. Toyota's lack of capital and low production volumes meant that the large, inflexible, limited-purpose press and die machines routine in U.S. auto plants were infeasible. As necessity is the mother of invention, Ohno's solution was to invent methods to move from rigid, highly specialized machines to flexible, multipurpose machines that could be changed over quickly and easily. Unlike Ford, which manufactured millions of one particular car model, Toyota was faced with low demand for a variety of different models. And because of the limited cash flow of this fledgling auto manufacturer, production was keyed to customer demand. There simply was no room for large inventories of finished goods waiting on dealer lots until a customer could be convinced to buy a particular automobile.

This unique set of circumstances led to small-batch production built to customer demand in short cycle times with minimal waste — the hallmarks of Lean Production.

12.4.1 GETTING STARTED WITH LEAN

The first step in taking a walk on the lean side is to conduct a *value stream analysis*.[12] This is defining the value of the product or service from the customer's perspective and answering the question of precisely why the customer is willing to allocate dollars to the particular product or service. A good value stream analysis specifies what the customer seeks in terms of requirements, price, and time. It must capture current customer requirements as well as potential future or emerging requirements.

The next step is to determine whether each stage of the current production process, or *value stream*, is adding value. Rework, approvals, storage, and queues are examples of non-valued-added activities. Value involves transforming inputs into outputs that a customer is willing to pay for — anything that does not add value needs to be eliminated from the production process. Ensuring only value-added steps is what lean is all about — eliminating all waste. The lean techniques, described in the following sections, are used to accomplish this goal.

12.4.2 CONTINUOUS-FLOW PRODUCTION

Continuous-flow production is synonymous with small-batch production. Ideally, one-piece flow is achieved, meaning a single product moves continuously through the production process using the principle of "make one, move one." Some of the advantages of continuous-flow production are:[1]

- Less work in process through having production workers and machines complete one piece before moving on to the next piece
- Reduced handling and transportation time
- Immediate feedback on any overlooked defect

To achieve continuous-flow production, the production line must be balanced among the different production steps. An everyday example of production line balancing is washing and drying clothes.[13] Typically the washing machine is faster than the dryer. The result is either a large amount of work-in-process inventory if you continue to run the washer or underutilized capacity of the washer while waiting for the dryer. To achieve line balance we must design and schedule the production flow so that each production step is matched to *takt* time. *Takt* is the German word for pace; in continuous-flow production, it is the overall production rate required for the entire system to match the customer demand. *Takt* time is calculated as:

Available work time per day / customer demand per day

If the customer requires 400 orders per day and the time available is two shifts of 480 min per day, then the *takt* time is:

(480 min x 2)/400 orders = 2.4 min per order

The *takt* time, or pace at which the entire operation should produce orders, is 0.42 orders per minute. Producing an order in less than 2.4 min results in excess inventory of finished goods; producing an order in more than 2.4 min results in not meeting the customer's demand for the orders. The goal of continuous-flow manufacturing is to balance production so that each production step produces 0.42 orders per minute.[12] (Note this is different from cycle time, which is the time from the customer order to the time the order is delivered.) It is unlikely that every process step can produce exactly to the *takt* time and thus, it is necessary to identify the steps in the production process that can be tied to the *takt* time. These are usually high-volume, repeatable steps. Ideally, these steps are close to final assembly and shipping so that the output of the process is more predictable and can be scheduled to the customer demand.

Continuous-flow production leads to the concept of *pull scheduling*, in which an order is not available for the next step in the production line until that step is ready to receive it. This is sometimes called a *supermarket pull system*, because it is similar to a supermarket in which goods are replaced on the shelves only after the customer has pulled them off for purchase. *Kanban* cards are used to signal the number of goods needed to replenish the shelves. Kanban is synonymous with *Just-in-Time (JIT)* inventory and production systems in which inventory is scheduled to arrive at the time of need, rather than stocked on site.[13]

The pull scheduling concept extends to the entire production system, in which goods are only produced after the customer has ordered them. Raw materials are received just in time in small batches on one end of the continuous-flow system, and the order is pulled through the production process through final assembly and shipped immediately to the customer. The cycle time from customer request to invoicing and payment receipt is appreciably shorter than in push scheduling, in which goods are produced according to marketing sales forecasts and then pushed onto the customer through intense sales efforts.

The design of Lean Production and Lean Servicing systems involves a number of tools and techniques:

- *Production cells.* To minimize the waste of transportation and moving, we replace the familiar assembly line with U-shaped production cells in which several machines are put together. Small, flexible, multipurpose machines arranged in a U shape in order of production are preferred over large, single-purpose machines that require extensive changeover time. Operators should be able to handle multiple tasks and machines while standing and moving freely within the production cell. This contrasts with "fat" production, in which operators predominantly sit at single-purpose machines, occasionally rising to replenish supplies. The small, flexible machines are stationed on moving casters rather than bolted to the floor, so that production sequencing can be changed quickly.[14,15]
- *Single-Minute Exchange of Dies (SMED).* SMED, which is a technique for quick changeovers of multipurpose machines, was developed by Shiego Shingo at Toyota to facilitate the use of continuous-flow and small-batch production. Small-batch production is incompatible with

large, single-purpose machines that require extensive changeover time to run different production lots. Think of a race car driver making a pit stop. The pit stop (changeover) time can make the difference between winning and losing. Changeovers add no value and can be eliminated through the use of small, single-purpose equipment, or minimized through SMED techniques. The design of the machine and the process supporting the changeovers is different under SMED than traditional production systems.[16]

- *Total Productive Maintenance.* Because Lean Production builds to customer request in a very short cycle of production, wasteful equipment down time must be minimized. Lean addresses this through TPM, in which equipment operators are responsible for preventive maintenance of their machines and have the authority to take corrective action if necessary. This intimate knowledge of and responsibility for their machines means the operators are more likely to understand what can be improved to make their jobs easier. Six Sigma reliability and scheduling techniques support the use of TPM.[1]

- *Emphasis on training and development.* Handling multiple process steps and maintaining their own equipment requires a highly skilled and cooperative workforce. Ongoing development and training of production operators is necessary. Workers are usually assigned to a team to ensure high flexibility in the production process. When a quality problem arises, team members use *jidoka*, meaning built-in quality, to stop the production process and immediately solve the problem. Calling on team members to assist in the process is common. The use of brainstorming and analytic tools to identify the root cause of the problem and develop solutions requires ongoing development and cooperation between management and production workers.[12]

- *Gemba kaizen. Gemba* means "close to the action," and in Lean Production, the production floor is where the action is. *Kaizen* (*kai* for change, *zen* for good) is used to describe continuous improvement activities. A *gemba kaizen* event is a commonsense approach to solving pressing quality issues through involving the production workers in a multiple-day off-the-floor session using lean and Six Sigma problem-solving tools. *Kaizen* is also the practice of ongoing incremental improvements by all involved in the production process.[14]

- *Visual control.* This technique makes information available at a glance to all involved in the process. A *visual factory* uses visual control extensively. Andon boards, on which a light or a sound is activated when a problem occurs, are a visual control tool. Another tool is to display control and performance charts, production status charts, and similar-type information. The idea is that everyone in the production process has the same level of information on the current status.[17]

- *The 5S's.* Similar to the concept of the visual factory is the 5S concept of rules of orderly, safe, and clean production processes. The Japanese terms for the 5S's are *Seiri, Seiton, Seison, Seiketsu,* and *Shitsuke.* The Western interpretation is as follows:[1]

- Sort and separate out all things that are unnecessary to the production process and eliminate them.
- Straighten and arrange essential production components in order for easy access.
- Scrub essential production equipment and keep the work areas clean.
- Stabilize the practice of good housekeeping by making cleanliness and orderliness a routine process.
- Sustain the 5S's as a way of life.

An example of both visual controls and 5S at work is placing tape on the floor around a movable machine so that everyone knows exactly where the machine should be placed. Remember that continuous-flow manufacturing requires small, flexible machines on casters, so the use of tape as a tool for temporary orderliness of equipment supports many lean practices.

- *Poka-Yoke.* Poka-Yoke is also known as mistake proofing; the idea is to design processes that are robust to mistakes. Examples of Poka-Yoke are built-in process controls or specialized equipment that act as sensors to identify mistakes.[18]
- *Cardboard engineering.* Cardboard engineering refers to the technique of modeling a potential production cell before actual implementation. It is used in a situation where an assembly line is being transformed to a U-shaped production cell, to elucidate the work flow issues involved before the actual changes are made.[19]

12.5 CASE STUDY CONTINUED

Allen and Anna work with the AP department managers, supervisors, and other associates to develop a value stream map. They start by identifying the CTQs for their customers and the business. The Atlanta call center randomly surveys approximately 300 Card One cardholders and 300 noncardholders through outbound calls to determine the CTQ attributes of the new card application process. The primary customer CTQ is a quick decision and quick mailing of the card once approved. Quick is defined as 5 days or less for decision making and mailing of the card within 1 day of the decision. The business CTQs are a sound credit risk decision, balancing risk vs. reward, based on relevant facts. One could argue that a sound credit risk decision is to offer no one credit (involves zero risk), but there is certainly no reward in that decision.

Figure 12.3 provides a value stream map for the new card application process, starting with receipt of the application and ending with the mailing of the credit card. The center processes approximately 9.4 million applications per year, or 36,000 per day. Card One has about a 33% approval rate, issuing approximately 11,160 cards each day.

The AP department runs 24 h per day, 5 days a week, with the volume evenly spread among the three shifts; this results in approximately 12,000 applications

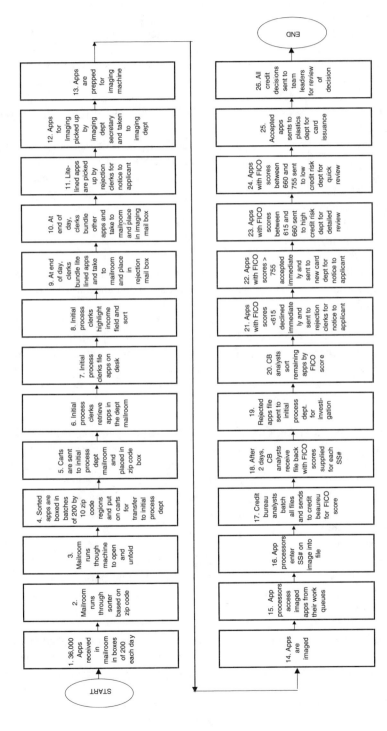

FIGURE 12.3 Value stream map for the new card application process.

entering the production process per shift, or 36,000 per day. The process starts in the mailroom, where the entire batch of 12,000 is received during the middle of each shift, so a new batch is started at the beginning of each 8-h shift. The mail arrives in boxes of 200 and each 200-lot batch runs through two machines: one machine to sort by zip code and another to open, unfold, and stack the applications. The applications are then stored again in boxes of 200 and loaded onto a cart for delivery to the second floor. The cart is transported once per hour in batches of 1,500 to 1,700 to the second floor mailroom and placed in the initial processing clerks' mailbox.

Allen and Anna do not notice anything unusual in the process until they come to the second floor. They are flabbergasted at what they observe. Huge boxes of nonworked applications are on the clerks' desk, although the clerks appear to be working quickly. The initial processing clerks' job is to review each application to determine if the applicant meets the minimum income criterion of $20,000 per year to be considered for a Card One credit card. The clerks quickly review and mark with a highlighting marker each application's income field in a process called lite-lining — the income is too "lite" to be considered for a card. They sort the applications into three stacks: lite-lined applications, applications with no income information, and applications that pass the minimum income criterion and will be processed later.

After spending a few hours estimating the work in process waiting to be lite-lined, Allen and Anna estimate a backlog of 2 days, or approximately 72,000 applications. This means that an application entering the mailroom today would not even be reviewed for another 3 days. They conclude that, if the bottleneck is here, the downstream process steps are sure to be waiting for applications to process. Unfortunately, they are wrong.

After the initial processing clerks completes their batches for a shift, the departmental support person bundles the lite-lined and missing information applications, takes them to the second floor mailroom, and places them in the rejection notification mailbox. There are about 2,200 lite-lined applications and 200 applications with missing information. The applications that meet the minimum income criterion are bundled and placed in the imaging department's mailbox for transportation to the first floor. At the end of the shift there is a batch of 8,500 placed in the imaging department's mailbox.

The next morning, the Lean Servicing duo of Allen and Anna begin their value charting exercise in the imaging department. The process is for the applications to be picked up at the end of the day by the mail clerks and transported to the imaging department's prep area where the prep clerks unbundle the applications and cull out damaged or folded ones. They try to smooth out applications in order to maximize the number of applications imaged; they set aside for special handling any applications they cannot prepare for the machine. The other applications are placed in stacks ready for imaging. About 1%, or 85, are set aside this morning, leaving a batch of approximately 8,400 to be imaged for the day shift. This is confusing to Allen because a shift has approximately 12,000 applications entering the production process each day, and historically the percentage of lite-lined and missing income applications runs at 20%, which leaves 9,600 per shift waiting for imaging. If 1%

is set aside for special handling, this should leave 9,500 per shift that are imaged. Then he remembers the backlog of applications waiting for lite-lining. Not only are they not catching up on the work-in-process inventory, the backlog is actually increasing. He makes a mental note to speak to the Director of Human Resources about adding more people to the initial processing area. As the lean duo make their way to the imaging machines, their spirits sag. Huge stacks of prepped applications waiting for imaging are piled up on every available desk and table in the small imaging room. The work in process is as great if not greater than the lite-lining backlog. The supervisor explains to Allen that the imaging machines have had a lot of down time lately due to a malfunctioning optical laser component. One of the two machines was down more than 50% of the time in the previous 2 days, sending work in process soaring. One of the imaging operators tells Allen that it is normal for the machines to be down at least 4 to 5 h per day for routine maintenance as well as malfunctions. The center is staffed to have both machines up and running a minimum of 21 h per machine per day. Allen quickly realizes that, under these conditions, they will never be able to clear out the backlog of applications; it will continue to grow. He begins to think he made a mistake in promising Harry that he could get the application processing time down to 3 days with less staff. What was he thinking? All of the Lean Servicing stuff sounded so right in class, but now he was unsure how to handle the fragmented departments, each with their own problems and growing backlogs.

From the imaging department, Allen and Anna move to the application processing department, where data entry clerks retrieve the image of the application on their desktop screen and enter the applicant's social security number (SSN) into a database to send to the credit bureau (CB). Compared to the previous two departments, the AP department seems to be operating smoothly. There does not appear to be a backlog in this department. The work pace is steady and fast.

The next department in the process, the credit bureau department, does not have any work in process; in fact, they appear to be waiting for work. After experiencing such heartache at seeing the huge work in process in the other departments, Allen welcomes the sight of a neat, clean, orderly, idle ... "Wait a minute," he thinks, "Am I nuts!?" There are several days' worth of work in process upstream, while downstream there is idle capacity. At least he can move some of the resources from this department to the ones that really need it. Or can he? The department with the excess capacity is responsible for sending the files from the AP data entry clerks to CBs for retrieval of the applicants' credit report, including their credit rating, known as the FICO score (from Fair Isaacs Company, who developed the scoring model). Two individuals work in this department, responsible primarily for ensuring the files are sent and received with all data intact. Upon receipt of a file back from the CB, the CB clerk sorts and segments the file by FICO score and routes the segmented file to one of four departments, depending on the score. Scores under 615 are immediately rejected; those files, representing approximately 15% of scored applications, are sent to the rejection notification department, where a letter is sent to the applicant informing of the decision. Files containing applications with scores above 755 (10% of the total) are sent to the plastics department on the first floor for issuance of a card. Because applicants with scores above 755 pose a very small credit risk, Card

One decided a couple of years ago to fully automate the process of issuing the card, with no human review. Sending the file to the plastics department sets in motion a process to automatically issue an acceptance letter and a second mailing with the card itself. Files containing applications with FICO scores between 615 and 660 are sent to the high-risk credit group for a detailed review. About 35% of scored applications are in this category. Files sent to the low-risk credit group are those with a score between 660 and 755 (about 40% of applications). They are reviewed to spot any unusual circumstances before a card is granted. Both credit risk groups require a team leader's review and approval before credit is granted, to ensure that an applicant who should not be granted credit is not mistakenly accepted. Approximately 1.2% of applications fall into this category and are caught by the team leader. Allen and Anna summarized their findings in Table 12.1.

They calculated the *takt* time of 2.4 sec by dividing 24 h per day by 36,000 applications per day. The total cycle time (from the perspective of the customer) to receive the new card was 20 days. The in-house cycle time was 13 days, up from the estimated 12 days just a couple of months before. Allen also suspected that applications requiring special processing were falling through the cracks — if they could not handle the routine processing, it was foolish to think that special processing was running smoothly.

Allen and Anna begin their Lean Servicing analysis by looking at the value stream in the application process. Remembering what they had learned in the Lean Servicing class, Allen and Anna look for the three value-added components in the value stream: (1) activities the customer is willing to pay for; (2) activities that add value through transformation of inputs into outputs; and (3) activities that are done right the first time. There are no "re" words: rework, review, or redundancies. The results are astonishing — they can hardly believe their eyes. Of the 13 in-house cycle time days, only about 2 days actually added value to the customer. Their new value stream map with the nonvalue activities crossed out is shown in Figure 12.4.

The next thing Allen remembers from his Lean Servicing class is to use continuous-flow manufacturing with small-batch flows wherever possible. He decides to analyze the production flow to try to better understand the impact of the large-batch processing the center is currently using. The first bottleneck caused by the large batches is in the initial processing department. Each clerk processes an entire shift's batch before passing it along to the next production activity, causing the entire production flow to be tied to their batch size. Any subsequent production step with faster processing time is forced to wait for the initial processing step to release their applications before they can start work. And the significant work-in-process inventory waiting to be worked in the initial processing department has caused the output of the entire center to slow down and be tied to the output and the backlog of this particular department.

In seeking areas of opportunity for continuous-flow manufacturing, Allen and Anna used the principles of Lean Servicing as well as common sense. They interviewed the employees on the front line to understand the issues with the current process and the barriers that kept the center from running at a steady, even pace. They researched the capacity of each production step to understand the bottlenecks, knowing that the entire output of the center was tied to the slowest process step.

TABLE 12.1
Summary of Value Stream Analysis Quantitative Data

Process Steps	No. of FTEs[a]	Batch Sizes	No. of Days in Department	Days of Work in Process	Days from Application 1 Entering Building	Days from Customer Sending Application	
Mailroom	4	1,500/hour	1			4	
Initial processing	23	12,000/shift		2	3	7	
Rejection notification	3	8,280/shift					
Imaging prep	2	9,600/shift	1		4	8	
Imaging	4	9,500/shift		3	7	11	
Application processing	10	9,500/shift	1		8	12	
Credit bureau	2	28,500/day	2		10	14	
Plastics	5	3,720/shift					
High credit risk	117	3,325/shift	1		11	15	
Low credit risk	80	3,800/shift					
Acceptance notification	3	2,770/shift	1		12	16	
Team leader review	8	2,800/shift	1		13	17	
Nonexempt support	10						
Exempt support staff	31					3	Days to mail
						20	Total cycle time in days for customer
Total staffing	302						

[a] Full-time equivalents.

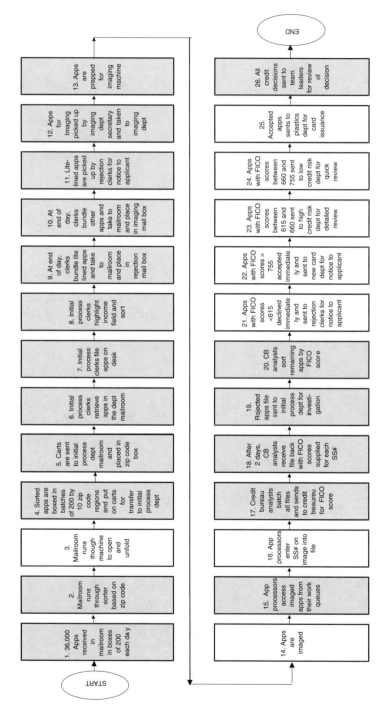

FIGURE 12.4 New value stream map with nonvalued activities darkened.

Once they identified initial application processing (lite-lining) as the bottleneck, they used their Six Sigma and TRIZ problem-solving tools to develop creative and innovative alternatives to the current process. Anna, who had worked at a high-tech manufacturing firm prior to Card One, was able to apply her knowledge of technology and expert credit decision systems to introduce a higher level of automation in the system, reducing or eliminating many of the routine and repetitive steps in the process. The expert credit system and other changes made in the process required substantially fewer employees. However, Card One was able to provide displaced employees with jobs in merchant operations, located on the third and fourth floors of the Dallas AP center. Merchant operations had just began to aggressively recruit new merchants while cultivating better relationships with their existing client base, thus requiring significant numbers of new employees. It was serendipity that at the same time merchant operations was looking to aggressively hire new employees, they had a pool of experienced Card One employees to hire from the first and second floor.

The specific changes resulting from adopting a Lean Servicing philosophy were as follows:

- The applications were no longer sorted by zip code. The zip code sorting was a carryover from a previous process that no longer applied.
- All applications were imaged before lite-lining rather than after.
- The imaging department was placed in the mailroom so that the distance between the mail machine and the image machine was minimal.
- A new high-speed imaging machine that imaged 2000 applications per hour replaced the older two machines. The extra lease cost on the new machine did not significantly increase costs relative to the two older machines.
- The new imaging machine was more robust to application damage than the previous ones, eliminating the need for prepping the applications before imaging.
- The new imaging machine also had optical character recognition capabilities, meaning it could "read" the data fields containing income and SSN. The success rate was 33%; thus 33% of the applications no longer required manual review of income and data entry of SSN.
- The imaging operators were trained on preventive maintenance of the machines and were given the responsibility of ensuring a high percentage of up time.
- The functions of the initial processing department were combined with the functions of the application processing department. The newly combined AP department was responsible for lite-lining as well as data entry of the applicants' SSNs. These tasks were combined into one job with a new standard of 220 units processed per hour (UPH). (A unit is one application.) Previously, the lite-line UPH was 220 and the SSN data entry UPH was 400. The volume sent to this department was reduced by 33% through use of the new imaging machine with OCR (optical character recognition) technology.

- Allen and Anna visited several CBs to understand their options relative to the 2-day turnaround from their current CB, which did not meet their customer or business CTQs. They were able to strike a deal with a new company that offered real-time reporting of the FICO score at a cost that was 20% higher than their current charge. However, they also decided to no longer order the full credit report if the FICO scores the CB supplied were less than the minimum of 615 or greater than the automatically accepted score of 755. Also, abbreviated credit reports would be ordered for applicants in the low-risk credit group. Reducing the number of full credit reports ordered more than offset the increased cost of real-time credit reporting.
- Along with the new real-time credit score reporting, they outsourced with a neural network company to provide an expert system to review and make credit decisions regarding the low- and high-risk credit applicants, based on the application information and their credit report. They were able to eliminate approximately half of the manual review and underwriting decisions that were currently performed in house by the two credit risk departments.
- Understanding the importance of continuous-flow manufacturing, and with the help of their Lean Servicing instructor, they were able to implement a small-batch production system for the mailroom and imaging functions. After these two production steps, the batch size was reduced to one for all the remaining steps. This means that each application is worked and then passed on immediately to the next processing step. They identified the remaining bottleneck in the production process as the high-risk credit department, because that was the slowest process step in the system, with a UPH of 12 (i.e., 5 min per application). This department was identified as the "pacemaker," because all other steps in the production process are tied to its production. They decided to staff this area to match the process output of the other production steps. Fortunately, the expert credit decision system relieved this department of about 50% of its workload. The low-risk credit department operated in parallel to the high-risk department, and because their output was not interdependent, they were flexibly staffed so that the rate of output of the two were the same. This way, the entire system's output was the same regardless of which credit decision department the application flowed through.

Figure 12.5 provides a summary of the new production statistics once the process was leveled. The small batch sizes dramatically decreased the time any one application was in the system, thus ensuring a shorter overall cycle time. Many applications were received and processed the same day through the use of the OCR technology, automated CB scoring, the expert credit decision system, and the automated plastics and letter generation systems.

The changes in the production sequence and the addition of new technologies reduced the average cycle time of in-house application processing from 13 days to 29 h. According to the Lean Servicing and Six Sigma philosophy of reporting on

Function	Input Daily Volume	Input Hourly Volume	Input Batch Size	Input Batches/ Day	Output Uph/ Fte	Output Min/App	Output Min/Day Required	Output H/Day Required	Output Productive Hours/FTE/ Day	Output FTEs Required to Handle Volume
Mailroom	36000	1500	250	144	600	0.10	3600	60	7.0	8.5
Imaging	36000	1500	250	144	2000	0.03	1080	18	7.0	2.5
App process	23976	999			220	0.27	6539	109	7.0	15.5
Credit bureau	24000	1000								
High risk	3570	149			12	5.00	17850	298	7.0	42.5
Low risk	5625	234			20	3.00	16875	281	7.0	40.5
Nonexempt support staff										7.5
Exempt support staff										27.0
Total staffing										144.0

Figure 12.5 Production statistics using Lean Servicing techniques.

the variation rather than the average, Allen changed the metrics of the center to report on a service-level measure rather than an average level. The new metric showed the center processed 90% of the applications within 35 h — better than the 3-day average turnaround Allen had promised Harry when he undertook the task of transforming into a lean servicer.

The AP center reduced their total full-time equivalents from 302 to 144, resulting in net savings of more than $3 million, even accounting for the increased cost of the imaging machine, the CB reports, and the expert credit systems. Figure 12.6 provides the new value stream analysis for the AP center.

It was not an easy transition. Many headaches and obstacles were encountered. The most challenging obstacles related to employee issues and the anxiety levels caused by the significant changes. Because of the training he received in the Lean Servicing class, Allen had expected this and coordinated with the Human Resources organizational development manager early on. The center implemented a change management effort parallel to the Lean Servicing transformation, reducing the opportunity for significant productivity losses due to the changes. The entire effort took a little more than 9 months, but the payback was almost immediate. Allen and Anna were celebrating the success of the effort with the center's employees when Harry

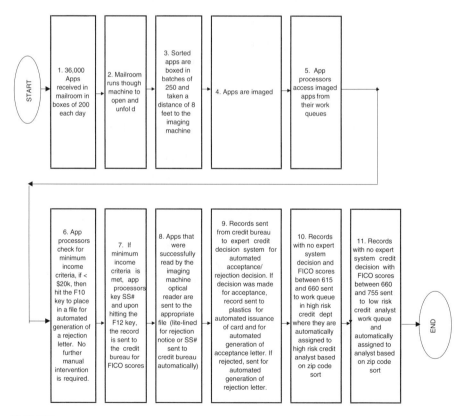

FIGURE 12.6 Value stream analysis using Lean Servicing techniques.

took Allen aside and informed him of a recent conversation he had had earlier in the week. As a result of the rapid expansion, it appeared that things at merchant operations were a little out of hand, with backlogs and increasing complaints from the Card One merchants. The executive vice-president of merchant operations had inquired about the possibility of borrowing Allen and Anna so that they could "work their magic." Allen laughed and commented to Harry that there was no magic involved; just a lot of sound techniques framed within a philosophy of offering the customer and the business true value through eliminating all waste in the process of meeting their CTQs. They made a toast to each other and to the concept of Lean Servicing.

REFERENCES

1. "Transferring Lean Manufacturing to Small Manufacturers: The Role of NIST-MEP," http://www.sbaer.uca.edu/Docs/proceedingsIII.htm, 2001.
2. Womack, J.P., Jones, D.T., and Roos, D., *The Machine that Changed the World: The Story of Lean Production*, HarperPerennial, New York, 1990. This book provided much of the information in this chapter.
3. Womack, J.P., Jones, D.T., and Roos, D., *The Machine that Changed the World: The Story of Lean Production*, HarperPerennial, New York, 1990, p. 26.
4. Ohno, T., *Toyota Production System — Beyond Large-Scale Production*, Productivity Press, Portland, OR, 1988.
5. Womack, J.P., Jones, D.T., and Roos, D., *The Machine that Changed the World: The Story of Lean Production*, HarperPerennial, New York, 1990, p. 27.
6. Womack, J.P., Jones, D.T., and Roos, D., *The Machine that Changed the World: The Story of Lean Production*, HarperPerennial, New York, 1990, p. 28.
7. Hicks, P.E., *Introduction to Industrial Engineering and Management Science*, McGraw-Hill, New York, 1977, p. 24.
8. Womack, J.P., Jones, D.T., and Roos, D., *The Machine that Changed the World: The Story of Lean Production*, HarperPerennial, New York, 1990, p. 39.
9. Womack, J.P., Jones, D.T., and Roos, D., *The Machine that Changed the World: The Story of Lean Production*, HarperPerennial, New York, 1990, p. 56.
10. "History of Lean: The Evolution of a New Manufacturing Paradigm," http://www.Optiprise.com/Company/Overview/History/Lean.htm, 2001.
11. Womack, J.P., Jones, D.T., and Roos, D., *The Machine that Changed the World: The Story of Lean Production*, HarperPerennial, New York, 1990, p. 57.
12. "Creating a Future State," http://www.Lean.org/Lean/Community/Resources/thinkers_start.cfm, 2001.
13. "Pull Scheduling/Just in Time," http://www.moresteam.com/lean/1608.cfm, 2001.
14. "Turning Japanese," http://www.autofieldguide.com/columns/article.cfm, 2001.
15. *Lean Manufacturing Advisor*, Volume 2, Number 10, March 2001.
16. "Quick Changeover (SMED)," http://www.moresteam.com/lean/1608.cfm, 2001.
17. From a lecture at Clemson University presented by Professor Iris D. Tommelein titled "Lean Thinking," February 21, 2001.
18. Breyfolge, F.W., *Implementing Six Sigma: Smarter Solutions Using Statistical Methods*, John Wiley & Sons, 1999, p. 551.
19. "Getting Lean Everywhere and Every Day," http://www.autofieldguide.com/columns/article.cfm, 2001.

Appendix A: Deming's 14 Points for Management

All anyone asks for is a chance to work with pride.

W. Edwards Deming*

1. Create constancy of purpose toward improvement of product and service, with the aim to become competitive and to stay in business, and to provide jobs.
2. Adopt the new philosophy. We are in a new economic age. Western management must awaken to the challenge, must learn their responsibilities, and take on leadership for change.
3. Cease dependence on inspection to achieve quality. Eliminate the need for inspection on a mass basis by building quality into the product in the first place.
4. End the practice of awarding business on the basis of price tag. Instead, minimize total cost. Move toward a single supplier for any one item, on a long-term relationship of loyalty and trust.
5. Improve constantly and forever the system of production and service, to improve quality and productivity, and thus constantly decrease costs.
6. Institute training on the job.
7. Institute leadership. The aim of supervision should be to help people and machines and gadgets to do a better job. Supervision of management is in need of overhaul, as well as supervision of production workers.
8. Drive out fear, so that everyone may work effectively for the company.
9. Break down barriers between departments. People in research, design, sales, and production must work as a team, to foresee problems of production and in use that may be encountered with the product or service.
10. Eliminate slogans, exhortations, and targets for the work force asking for zero defects and new levels of productivity. Such exhortations only create adversarial relationships, as the bulk of the causes of low quality and low productivity belong to the system and thus lie beyond the power of the work force.

---

* http://www.deming.org.

11. a. Eliminate work standards (quotas) on the factory floor. Substitute leadership.

    b. Eliminate management by objective. Eliminate management by numbers, numerical goals. Substitute leadership.

12. a. Remove barriers that rob the hourly worker of his right to pride of workmanship. The responsibility of supervisors must be changed from sheer numbers to quality.

    b. Remove barriers that rob people in management and in engineering of their right to pride of workmanship. This means, inter alia, abolishment of the annual or merit rating and of management by objective.

13. Institute a vigorous program of education and self-improvement.

14. Put everybody in the company to work to accomplish the transformation. The transformation is everybody's job.

Appendix B: Statistical Tables Used for Six Sigma

B.1 AREA UNDER THE NORMAL CURVE

Table B.1 contains the area under the standard normal curve from 0 to z. It can be used to compute the cumulative distribution function values for the standard normal distribution.

Table B.1 utilizes the symmetry of the normal distribution, so what in fact is given is

$$P[0 \leq x \leq |a|]$$

where a is the value of interest. This is demonstrated in the graph below for $a = 0.5$. The shaded area of the curve represents the probability that x is between 0 and a. z

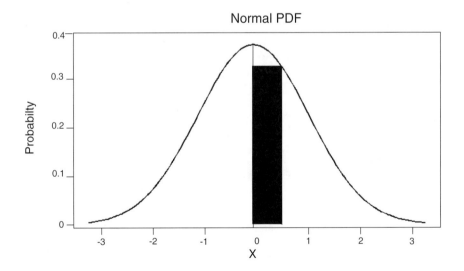

Normal PDF

TABLE B.1
Critical Values for the Normal Distribution

Z	0.00	0.01	0.02	0.03	0.04	0.05	0.06	0.07	0.08	0.09
0.0	0.5000	0.5040	0.5080	0.5120	0.5160	0.5199	0.5239	0.5279	0.5319	0.5359
0.1	0.5398	0.5438	0.5478	0.5517	0.5557	0.5596	0.5636	0.5675	0.5714	0.5753
0.2	0.5793	0.5832	0.5871	0.5910	0.5948	0.5987	0.6026	0.6064	0.6103	0.6141
0.3	0.6179	0.6217	0.6255	0.6293	0.6331	0.6368	0.6406	0.6443	0.6480	0.6517
0.4	0.6554	0.6591	0.6628	0.6664	0.6700	0.6736	0.6772	0.6808	0.6844	0.6879
0.5	0.6915	0.6950	0.6985	0.7019	0.7054	0.7088	0.7123	0.7157	0.7190	0.7224
0.6	0.7257	0.7291	0.7324	0.7357	0.7389	0.7422	0.7454	0.7486	0.7517	0.7549
0.7	0.7580	0.7611	0.7642	0.7673	0.7704	0.7734	0.7764	0.7794	0.7823	0.7852
0.8	0.7881	0.7910	0.7939	0.7967	0.7995	0.8023	0.8051	0.8078	0.8106	0.8133
0.9	0.8159	0.8186	0.8212	0.8238	0.8264	0.8289	0.8315	0.8340	0.8365	0.8389
1.0	0.8413	0.8438	0.8461	0.8485	0.8508	0.8531	0.8554	0.8577	0.8599	0.8621
1.1	0.8643	0.8665	0.8686	0.8708	0.8729	0.8749	0.8770	0.8790	0.8810	0.8830
1.2	0.8849	0.8869	0.8888	0.8907	0.8925	0.8944	0.8962	0.8980	0.8997	0.9015
1.3	0.9032	0.9049	0.9066	0.9082	0.9099	0.9115	0.9131	0.9147	0.9162	0.9177
1.4	0.9192	0.9207	0.9222	0.9236	0.9251	0.9265	0.9279	0.9292	0.9306	0.9319
1.5	0.9332	0.9345	0.9357	0.9370	0.9382	0.9394	0.9406	0.9418	0.9429	0.9441

1.6	—	0.9452	0.9463	0.9474	0.9484	0.9495	0.9505	0.9515	0.9525	0.9535	0.9545
1.7	—	0.9554	0.9564	0.9573	0.9582	0.9591	0.9599	0.9608	0.9616	0.9625	0.9633
1.8	—	0.9641	0.9649	0.9656	0.9664	0.9671	0.9678	0.9686	0.9693	0.9699	0.9706
1.9	—	0.9713	0.9719	0.9726	0.9732	0.9738	0.9744	0.9750	0.9756	0.9761	0.9767
2.0	—	0.9772	0.9778	0.9783	0.9788	0.9793	0.9798	0.9803	0.9808	0.9812	0.9817
2.1	—	0.9821	0.9826	0.9830	0.9834	0.9838	0.9842	0.9846	0.9850	0.9854	0.9857
2.2	—	0.9861	0.9864	0.9868	0.9871	0.9875	0.9878	0.9881	0.9884	0.9887	0.9890
2.3	—	0.9893	0.9896	0.9898	0.9901	0.9904	0.9906	0.9909	0.9911	0.9913	0.9916
2.4	—	0.9918	0.9920	0.9922	0.9925	0.9927	0.9929	0.9931	0.9932	0.9934	0.9936
2.5	—	0.9938	0.9940	0.9941	0.9943	0.9945	0.9946	0.9948	0.9949	0.9951	0.9952
2.6	—	0.9953	0.9955	0.9956	0.9957	0.9959	0.9960	0.9961	0.9962	0.9963	0.9964
2.7	—	0.9965	0.9966	0.9967	0.9968	0.9969	0.9970	0.9971	0.9972	0.9973	0.9974
2.8	—	0.9974	0.9975	0.9976	0.9977	0.9977	0.9978	0.9979	0.9979	0.9980	0.9981
2.9	—	0.9981	0.9982	0.9982	0.9983	0.9984	0.9984	0.9985	0.9985	0.9986	0.9986
3.0	—	0.9987	0.9987	0.9987	0.9988	0.9988	0.9989	0.9989	0.9989	0.9990	0.9990

Adapted from the *National Institute of Standards and Technology Engineering Statistics Handbook.*

B.2 UPPER CRITICAL VALUES OF THE STUDENT'S
t DISTRIBUTION

B.2.1 HOW TO USE THIS TABLE

Table B.2 contains the upper critical values of the student's *t* distribution. The upper critical values are computed using the percent point function. Due to the symmetry of the *t* distribution, this table can be used for both one-sided (lower and upper) and two-sided tests using the appropriate value of α.

The significance level, α, is demonstrated with the graph below, which plots a *t* distribution with 10 degrees of freedom. The most commonly used significance level is α = .05. For a two-sided test, we compute the percent point function at α/2 (.025). If the absolute value of the test statistic is greater than the upper critical value (.025), then we reject the null hypothesis. Due to the symmetry of the *t* distribution, we tabulate only the upper critical values in Table B.2.

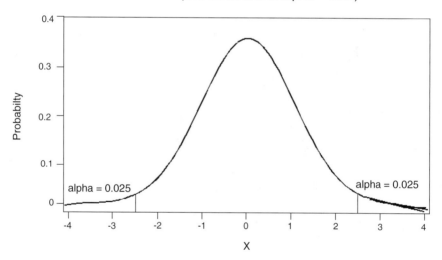

t PDF (two-sided test at alpha = 0.05)

Given a specified value for α:

For a two-sided test, find the column corresponding to α/2 and reject the null hypothesis if the absolute value of the test statistic is greater than the value of $t_{\alpha/2}$ in Table B.2.

For an upper one-sided test, find the column corresponding to α and reject the null hypothesis if the test statistic is greater than the Table B.2 value.

For a lower one-sided test, find the column corresponding to α and reject the null hypothesis if the test statistic is less than the negative of the Table B.2 value.

TABLE B.2
Upper Critical Values of Student's t Distribution with Degrees of Freedom (v)

Probability of Exceeding the Critical Value

v	0.10	0.05	0.025	0.01	0.001
1	3.078	6.314	12.706	31.821	318.313
2	1.886	2.920	4.303	6.965	22.327
3	1.638	2.353	3.182	4.541	10.215
4	1.533	2.132	2.776	3.747	7.173
5	1.476	2.015	2.571	3.365	5.893
6	1.440	1.943	2.447	3.143	5.208
7	1.415	1.895	2.365	2.998	4.782
8	1.397	1.860	2.306	2.896	4.499
9	1.383	1.833	2.262	2.821	4.296
10	1.372	1.812	2.228	2.764	4.143
11	1.363	1.796	2.201	2.718	4.024
12	1.356	1.782	2.179	2.681	3.929
13	1.350	1.771	2.160	2.650	3.852
14	1.345	1.761	2.145	2.624	3.787
15	1.341	1.753	2.131	2.602	3.733
16	1.337	1.746	2.120	2.583	3.686
17	1.333	1.740	2.110	2.567	3.646
18	1.330	1.734	2.101	2.552	3.610
19	1.328	1.729	2.093	2.539	3.579
20	1.325	1.725	2.086	2.528	3.552
21	1.323	1.721	2.080	2.518	3.527
22	1.321	1.717	2.074	2.508	3.505
23	1.319	1.714	2.069	2.500	3.485
24	1.318	1.711	2.064	2.492	3.467
25	1.316	1.708	2.060	2.485	3.450
26	1.315	1.706	2.056	2.479	3.435
27	1.314	1.703	2.052	2.473	3.421
28	1.313	1.701	2.048	2.467	3.408
29	1.311	1.699	2.045	2.462	3.396
30	1.310	1.697	2.042	2.457	3.385
31	1.309	1.696	2.040	2.453	3.375
32	1.309	1.694	2.037	2.449	3.365
33	1.308	1.692	2.035	2.445	3.356
34	1.307	1.691	2.032	2.441	3.348
35	1.306	1.690	2.030	2.438	3.340
36	1.306	1.688	2.028	2.434	3.333
37	1.305	1.687	2.026	2.431	3.326
38	1.304	1.686	2.024	2.429	3.319
39	1.304	1.685	2.023	2.426	3.313
40	1.303	1.684	2.021	2.423	3.307
41	1.303	1.683	2.020	2.421	3.301

TABLE B.2 (continued)
Upper Critical Values of Student's *t* Distribution with Degrees
of Freedom (ν)

Probability of Exceeding the Critical Value

ν	0.10	0.05	0.025	0.01	0.001
42	1.302	1.682	2.018	2.418	3.296
43	1.302	1.681	2.017	2.416	3.291
44	1.301	1.680	2.015	2.414	3.286
45	1.301	1.679	2.014	2.412	3.281
46	1.300	1.679	2.013	2.410	3.277
47	1.300	1.678	2.012	2.408	3.273
48	1.299	1.677	2.011	2.407	3.269
49	1.299	1.677	2.010	2.405	3.265
50	1.299	1.676	2.009	2.403	3.261
51	1.298	1.675	2.008	2.402	3.258
52	1.298	1.675	2.007	2.400	3.255
53	1.298	1.674	2.006	2.399	3.251
54	1.297	1.674	2.005	2.397	3.248
55	1.297	1.673	2.004	2.396	3.245
56	1.297	1.673	2.003	2.395	3.242
57	1.297	1.672	2.002	2.394	3.239
58	1.296	1.672	2.002	2.392	3.237
59	1.296	1.671	2.001	2.391	3.234
60	1.296	1.671	2.000	2.390	3.232
61	1.296	1.670	2.000	2.389	3.229
62	1.295	1.670	1.999	2.388	3.227
63	1.295	1.669	1.998	2.387	3.225
64	1.295	1.669	1.998	2.386	3.223
65	1.295	1.669	1.997	2.385	3.220
66	1.295	1.668	1.997	2.384	3.218
67	1.294	1.668	1.996	2.383	3.216
68	1.294	1.668	1.995	2.382	3.214
69	1.294	1.667	1.995	2.382	3.213
70	1.294	1.667	1.994	2.381	3.211
71	1.294	1.667	1.994	2.380	3.209
72	1.293	1.666	1.993	2.379	3.207
73	1.293	1.666	1.993	2.379	3.206
74	1.293	1.666	1.993	2.378	3.204
75	1.293	1.665	1.992	2.377	3.202
76	1.293	1.665	1.992	2.376	3.201
77	1.293	1.665	1.991	2.376	3.199
78	1.292	1.665	1.991	2.375	3.198
79	1.292	1.664	1.990	2.374	3.197
80	1.292	1.664	1.990	2.374	3.195
81	1.292	1.664	1.990	2.373	3.194
82	1.292	1.664	1.989	2.373	3.193

TABLE B.2 (continued)
Upper Critical Values of Student's *t* Distribution with Degrees of Freedom (ν)

Probability of Exceeding the Critical Value

ν	0.10	0.05	0.025	0.01	0.001
83	1.292	1.663	1.989	2.372	3.191
84	1.292	1.663	1.989	2.372	3.190
85	1.292	1.663	1.988	2.371	3.189
86	1.291	1.663	1.988	2.370	3.188
87	1.291	1.663	1.988	2.370	3.187
88	1.291	1.662	1.987	2.369	3.185
89	1.291	1.662	1.987	2.369	3.184
90	1.291	1.662	1.987	2.368	3.183
91	1.291	1.662	1.986	2.368	3.182
92	1.291	1.662	1.986	2.368	3.181
93	1.291	1.661	1.986	2.367	3.180
94	1.291	1.661	1.986	2.367	3.179
95	1.291	1.661	1.985	2.366	3.178
96	1.290	1.661	1.985	2.366	3.177
97	1.290	1.661	1.985	2.365	3.176
98	1.290	1.661	1.984	2.365	3.175
99	1.290	1.660	1.984	2.365	3.175
100	1.290	1.660	1.984	2.364	3.174
∞	1.282	1.645	1.960	2.326	3.090

Adapted from the *National Institute of Standards and Technology Engineering Statistics Handbook*.

B.3 UPPER CRITICAL VALUES OF THE *F* DISTRIBUTION

B.3.1 How to Use This Table

Table B.3 contains the upper critical values of the F distribution. This table is used for one-sided F tests at the $\alpha = .05, .10$, and $.01$ levels.

More specifically, a test statistic is computed with v_1 and v_2 degrees of freedom, and the result is compared to Table B.3. For a one-sided test, the null hypothesis is rejected when the test statistic is greater than the tabled value. This is demonstrated with the graph of an F distribution with $v_1 = 10$ and $v_2 = 10$. The shaded area of the graph indicates the rejection region at the α significance level. Because this is a one-sided test, we have α probability in the upper tail of exceeding the critical value and zero in the lower tail. Because the F distribution is asymmetric, a two-sided test requires a set of tables (not included here) that contain the rejection regions for both the lower and upper tails.

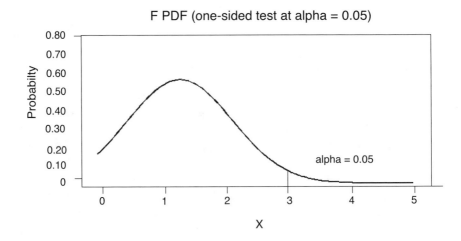

TABLE B.3
Upper Critical Values of the F Distribution for v_1 Numerator Degrees of Freedom and v_2 Denominator Degrees of Freedom

5% Significance Level
$F_{.05}(v_1, v_2)$

v_2	v_1									
	1	2	3	4	5	6	7	8	9	10
1	161.448	199.500	215.707	224.583	230.162	233.986	236.768	238.882	240.543	241.882
2	18.513	19.000	19.164	19.247	19.296	19.330	19.353	19.371	19.385	19.396
3	10.128	9.552	9.277	9.117	9.013	8.941	8.887	8.845	8.812	8.786
4	7.709	6.944	6.591	6.388	6.256	6.163	6.094	6.041	5.999	5.964
5	6.608	5.786	5.409	5.192	5.050	4.950	4.876	4.818	4.772	4.735
6	5.987	5.143	4.757	4.534	4.387	4.284	4.207	4.147	4.099	4.060
7	5.591	4.737	4.347	4.120	3.972	3.866	3.787	3.726	3.677	3.637
8	5.318	4.459	4.066	3.838	3.687	3.581	3.500	3.438	3.388	3.347
9	5.117	4.256	3.863	3.633	3.482	3.374	3.293	3.230	3.179	3.137
10	4.965	4.103	3.708	3.478	3.326	3.217	3.135	3.072	3.020	2.978
11	4.844	3.982	3.587	3.357	3.204	3.095	3.012	2.948	2.896	2.854
12	4.747	3.885	3.490	3.259	3.106	2.996	2.913	2.849	2.796	2.753
13	4.667	3.806	3.411	3.179	3.025	2.915	2.832	2.767	2.714	2.671
14	4.600	3.739	3.344	3.112	2.958	2.848	2.764	2.699	2.646	2.602
15	4.543	3.682	3.287	3.056	2.901	2.790	2.707	2.641	2.588	2.544
16	4.494	3.634	3.239	3.007	2.852	2.741	2.657	2.591	2.538	2.494
17	4.451	3.592	3.197	2.965	2.810	2.699	2.614	2.548	2.494	2.450
18	4.414	3.555	3.160	2.928	2.773	2.661	2.577	2.510	2.456	2.412
19	4.381	3.522	3.127	2.895	2.740	2.628	2.544	2.477	2.423	2.378

TABLE B.3 (continued)
Upper Critical Values of the F Distribution for v_1 Numerator Degrees of Freedom and v_2 Denominator Degrees of Freedom

5% Significance Level
$F_{.05}(v_1, v_2)$

v_2	v_1 1	2	3	4	5	6	7	8	9	10
20	4.351	3.493	3.098	2.866	2.711	2.599	2.514	2.447	2.393	2.348
21	4.325	3.467	3.072	2.840	2.685	2.573	2.488	2.420	2.366	2.321
22	4.301	3.443	3.049	2.817	2.661	2.549	2.464	2.397	2.342	2.297
23	4.279	3.422	3.028	2.796	2.640	2.528	2.442	2.375	2.320	2.275
24	4.260	3.403	3.009	2.776	2.621	2.508	2.423	2.355	2.300	2.255
25	4.242	3.385	2.991	2.759	2.603	2.490	2.405	2.337	2.282	2.236
26	4.225	3.369	2.975	2.743	2.587	2.474	2.388	2.321	2.265	2.220
27	4.210	3.354	2.960	2.728	2.572	2.459	2.373	2.305	2.250	2.204
28	4.196	3.340	2.947	2.714	2.558	2.445	2.359	2.291	2.236	2.190
29	4.183	3.328	2.934	2.701	2.545	2.432	2.346	2.278	2.223	2.177
30	4.171	3.316	2.922	2.690	2.534	2.421	2.334	2.266	2.211	2.165
31	4.160	3.305	2.911	2.679	2.523	2.409	2.323	2.255	2.199	2.153
32	4.149	3.295	2.901	2.668	2.512	2.399	2.313	2.244	2.189	2.142
33	4.139	3.285	2.892	2.659	2.503	2.389	2.303	2.235	2.179	2.133
34	4.130	3.276	2.883	2.650	2.494	2.380	2.294	2.225	2.170	2.123
35	4.121	3.267	2.874	2.641	2.485	2.372	2.285	2.217	2.161	2.114
36	4.113	3.259	2.866	2.634	2.477	2.364	2.277	2.209	2.153	2.106
37	4.105	3.252	2.859	2.626	2.470	2.356	2.270	2.201	2.145	2.098
38	4.098	3.245	2.852	2.619	2.463	2.349	2.262	2.194	2.138	2.091

39	2.084	2.131	2.187	2.255	2.342	2.456	2.612	2.845	3.238	4.091
40	2.077	2.124	2.180	2.249	2.336	2.449	2.606	2.839	3.232	4.085
41	2.071	2.118	2.174	2.243	2.330	2.443	2.600	2.833	3.226	4.079
42	2.065	2.112	2.168	2.237	2.324	2.438	2.594	2.827	3.220	4.073
43	2.059	2.106	2.163	2.232	2.318	2.432	2.589	2.822	3.214	4.067
44	2.054	2.101	2.157	2.226	2.313	2.427	2.584	2.816	3.209	4.062
45	2.049	2.096	2.152	2.221	2.308	2.422	2.579	2.812	3.204	4.057
46	2.044	2.091	2.147	2.216	2.304	2.417	2.574	2.807	3.200	4.052
47	2.039	2.086	2.143	2.212	2.299	2.413	2.570	2.802	3.195	4.047
48	2.035	2.082	2.138	2.207	2.295	2.409	2.565	2.798	3.191	4.043
49	2.030	2.077	2.134	2.203	2.290	2.404	2.561	2.794	3.187	4.038
50	2.026	2.073	2.130	2.199	2.286	2.400	2.557	2.790	3.183	4.034
51	2.022	2.069	2.126	2.195	2.283	2.397	2.553	2.786	3.179	4.030
52	2.018	2.066	2.122	2.192	2.279	2.393	2.550	2.783	3.175	4.027
53	2.015	2.062	2.119	2.188	2.275	2.389	2.546	2.779	3.172	4.023
54	2.011	2.059	2.115	2.185	2.272	2.386	2.543	2.776	3.168	4.020
55	2.008	2.055	2.112	2.181	2.269	2.383	2.540	2.773	3.165	4.016
56	2.005	2.052	2.109	2.178	2.266	2.380	2.537	2.769	3.162	4.013
57	2.001	2.049	2.106	2.175	2.263	2.377	2.534	2.766	3.159	4.010
58	1.998	2.046	2.103	2.172	2.260	2.374	2.531	2.764	3.156	4.007
59	1.995	2.043	2.100	2.169	2.257	2.371	2.528	2.761	3.153	4.004
60	1.993	2.040	2.097	2.167	2.254	2.368	2.525	2.758	3.150	4.001
61	1.990	2.037	2.094	2.164	2.251	2.366	2.523	2.755	3.148	3.998
62	1.987	2.035	2.092	2.161	2.249	2.363	2.520	2.753	3.145	3.996
63	1.985	2.032	2.089	2.159	2.246	2.361	2.518	2.751	3.143	3.993
64	1.982	2.030	2.087	2.156	2.244	2.358	2.515	2.748	3.140	3.991
65	1.980	2.027	2.084	2.154	2.242	2.356	2.513	2.746	3.138	3.989
66	1.977	2.025	2.082	2.152	2.239	2.354	2.511	2.744	3.136	3.986
67	1.975	2.023	2.080	2.150	2.237	2.352	2.509	2.742	3.134	3.984
68	1.973	2.021	2.078	2.148	2.235	2.350	2.507	2.740	3.132	3.982

TABLE B.3 (continued)
Upper Critical Values of the F Distribution for v_1 Numerator Degrees of Freedom and v_2 Denominator Degrees of Freedom

5% Significance Level
$F_{.05}(v_1, v_2)$

v_2	v_1									
	1	2	3	4	5	6	7	8	9	10
69	3.980	3.130	2.737	2.505	2.348	2.233	2.145	2.076	2.019	1.971
70	3.978	3.128	2.736	2.503	2.346	2.231	2.143	2.074	2.017	1.969
71	3.976	3.126	2.734	2.501	2.344	2.229	2.142	2.072	2.015	1.967
72	3.974	3.124	2.732	2.499	2.342	2.227	2.140	2.070	2.013	1.965
73	3.972	3.122	2.730	2.497	2.340	2.226	2.138	2.068	2.011	1.963
74	3.970	3.120	2.728	2.495	2.338	2.224	2.136	2.066	2.009	1.961
75	3.968	3.119	2.727	2.494	2.337	2.222	2.134	2.064	2.007	1.959
76	3.967	3.117	2.725	2.492	2.335	2.220	2.133	2.063	2.006	1.958
77	3.965	3.115	2.723	2.490	2.333	2.219	2.131	2.061	2.004	1.956
78	3.963	3.114	2.722	2.489	2.332	2.217	2.129	2.059	2.002	1.954
79	3.962	3.112	2.720	2.487	2.330	2.216	2.128	2.058	2.001	1.953
80	3.960	3.111	2.719	2.486	2.329	2.214	2.126	2.056	1.999	1.951
81	3.959	3.109	2.717	2.484	2.327	2.213	2.125	2.055	1.998	1.950
82	3.957	3.108	2.716	2.483	2.326	2.211	2.123	2.053	1.996	1.948
83	3.956	3.107	2.715	2.482	2.324	2.210	2.122	2.052	1.995	1.947
84	3.955	3.105	2.713	2.480	2.323	2.209	2.121	2.051	1.993	1.945
85	3.953	3.104	2.712	2.479	2.322	2.207	2.119	2.049	1.992	1.944
86	3.952	3.103	2.711	2.478	2.321	2.206	2.118	2.048	1.991	1.943
87	3.951	3.101	2.709	2.476	2.319	2.205	2.117	2.047	1.989	1.941

v_1

v_2	11	12	13	14	15	16	17	18	19	20
88	3.949	3.100	2.708	2.475	2.318	2.203	2.115	2.045	1.988	1.940
89	3.948	3.099	2.707	2.474	2.317	2.202	2.114	2.044	1.987	1.939
90	3.947	3.098	2.706	2.473	2.316	2.201	2.113	2.043	1.986	1.938
91	3.946	3.097	2.705	2.472	2.315	2.200	2.112	2.042	1.984	1.936
92	3.945	3.095	2.704	2.471	2.313	2.199	2.111	2.041	1.983	1.935
93	3.943	3.094	2.703	2.470	2.312	2.198	2.110	2.040	1.982	1.934
94	3.942	3.093	2.701	2.469	2.311	2.197	2.109	2.038	1.981	1.933
95	3.941	3.092	2.700	2.467	2.310	2.196	2.108	2.037	1.980	1.932
96	3.940	3.091	2.699	2.466	2.309	2.195	2.106	2.036	1.979	1.931
97	3.939	3.090	2.698	2.465	2.308	2.194	2.105	2.035	1.978	1.930
98	3.938	3.089	2.697	2.465	2.307	2.193	2.104	2.034	1.977	1.929
99	3.937	3.088	2.696	2.464	2.306	2.192	2.103	2.033	1.976	1.928
100	3.936	3.087	2.696	2.463	2.305	2.191	2.103	2.032	1.975	1.927

v_1

v_2	11	12	13	14	15	16	17	18	19	20
1	242.983	243.906	244.690	245.364	245.950	246.464	246.918	247.323	247.686	248.013
2	19.405	19.413	19.419	19.424	19.429	19.433	19.437	19.440	19.443	19.446
3	8.763	8.745	8.729	8.715	8.703	8.692	8.683	8.675	8.667	8.660
4	5.936	5.912	5.891	5.873	5.858	5.844	5.832	5.821	5.811	5.803
5	4.704	4.678	4.655	4.636	4.619	4.604	4.590	4.579	4.568	4.558
6	4.027	4.000	3.976	3.956	3.938	3.922	3.908	3.896	3.884	3.874
7	3.603	3.575	3.550	3.529	3.511	3.494	3.480	3.467	3.455	3.445
8	3.313	3.284	3.259	3.237	3.218	3.202	3.187	3.173	3.161	3.150
9	3.102	3.073	3.048	3.025	3.006	2.989	2.974	2.960	2.948	2.936
10	2.943	2.913	2.887	2.865	2.845	2.828	2.812	2.798	2.785	2.774
11	2.818	2.788	2.761	2.739	2.719	2.701	2.685	2.671	2.658	2.646
12	2.717	2.687	2.660	2.637	2.617	2.599	2.583	2.568	2.555	2.544
13	2.635	2.604	2.577	2.554	2.533	2.515	2.499	2.484	2.471	2.459

TABLE B.3 (continued)
Upper Critical Values of the F Distribution for ν_1 Numerator Degrees of Freedom and ν_2 Denominator Degrees of Freedom

5% Significance Level
$F_{.05}(\nu_1,\nu_2)$

ν_1

ν_2	11	12	13	14	15	16	17	18	19	20
14	2.565	2.534	2.507	2.484	2.463	2.445	2.428	2.413	2.400	2.388
15	2.507	2.475	2.448	2.424	2.403	2.385	2.368	2.353	2.340	2.328
16	2.456	2.425	2.397	2.373	2.352	2.333	2.317	2.302	2.288	2.276
17	2.413	2.381	2.353	2.329	2.308	2.289	2.272	2.257	2.243	2.230
18	2.374	2.342	2.314	2.290	2.269	2.250	2.233	2.217	2.203	2.191
19	2.340	2.308	2.280	2.256	2.234	2.215	2.198	2.182	2.168	2.155
20	2.310	2.278	2.250	2.225	2.203	2.184	2.167	2.151	2.137	2.124
21	2.283	2.250	2.222	2.197	2.176	2.156	2.139	2.123	2.109	2.096
22	2.259	2.226	2.198	2.173	2.151	2.131	2.114	2.098	2.084	2.071
23	2.236	2.204	2.175	2.150	2.128	2.109	2.091	2.075	2.061	2.048
24	2.216	2.183	2.155	2.130	2.108	2.088	2.070	2.054	2.040	2.027
25	2.198	2.165	2.136	2.111	2.089	2.069	2.051	2.035	2.021	2.007
26	2.181	2.148	2.119	2.094	2.072	2.052	2.034	2.018	2.003	1.990
27	2.166	2.132	2.103	2.078	2.056	2.036	2.018	2.002	1.987	1.974
28	2.151	2.118	2.089	2.064	2.041	2.021	2.003	1.987	1.972	1.959
29	2.138	2.104	2.075	2.050	2.027	2.007	1.989	1.973	1.958	1.945
30	2.126	2.092	2.063	2.037	2.015	1.995	1.976	1.960	1.945	1.932
31	2.114	2.080	2.051	2.026	2.003	1.983	1.965	1.948	1.933	1.920
32	2.103	2.070	2.040	2.015	1.992	1.972	1.953	1.937	1.922	1.908

33	2.093	2.060	2.030	2.004	1.982	1.961	1.943	1.926	1.911	1.898
34	2.084	2.050	2.021	1.995	1.972	1.952	1.933	1.917	1.902	1.888
35	2.075	2.041	2.012	1.986	1.963	1.942	1.924	1.907	1.892	1.878
36	2.067	2.033	2.003	1.977	1.954	1.934	1.915	1.899	1.883	1.870
37	2.059	2.025	1.995	1.969	1.946	1.926	1.907	1.890	1.875	1.861
38	2.051	2.017	1.988	1.962	1.939	1.918	1.899	1.883	1.867	1.853
39	2.044	2.010	1.981	1.954	1.931	1.911	1.892	1.875	1.860	1.846
40	2.038	2.003	1.974	1.948	1.924	1.904	1.885	1.868	1.853	1.839
41	2.031	1.997	1.967	1.941	1.918	1.897	1.879	1.862	1.846	1.832
42	2.025	1.991	1.961	1.935	1.912	1.891	1.872	1.855	1.840	1.826
43	2.020	1.985	1.955	1.929	1.906	1.885	1.866	1.849	1.834	1.820
44	2.014	1.980	1.950	1.924	1.900	1.879	1.861	1.844	1.828	1.814
45	2.009	1.974	1.945	1.918	1.895	1.874	1.855	1.838	1.823	1.808
46	2.004	1.969	1.940	1.913	1.890	1.869	1.850	1.833	1.817	1.803
47	1.999	1.965	1.935	1.908	1.885	1.864	1.845	1.828	1.812	1.798
48	1.995	1.960	1.930	1.904	1.880	1.859	1.840	1.823	1.807	1.793
49	1.990	1.956	1.926	1.899	1.876	1.855	1.836	1.819	1.803	1.789
50	1.986	1.952	1.921	1.895	1.871	1.850	1.831	1.814	1.798	1.784
51	1.982	1.947	1.917	1.891	1.867	1.846	1.827	1.810	1.794	1.780
52	1.978	1.944	1.913	1.887	1.863	1.842	1.823	1.806	1.790	1.776
53	1.975	1.940	1.910	1.883	1.859	1.838	1.819	1.802	1.786	1.772
54	1.971	1.936	1.906	1.879	1.856	1.835	1.816	1.798	1.782	1.768
55	1.968	1.933	1.903	1.876	1.852	1.831	1.812	1.795	1.779	1.764
56	1.964	1.930	1.899	1.873	1.849	1.828	1.809	1.791	1.775	1.761
57	1.961	1.926	1.896	1.869	1.846	1.824	1.805	1.788	1.772	1.757
58	1.958	1.923	1.893	1.866	1.842	1.821	1.802	1.785	1.769	1.754
59	1.955	1.920	1.890	1.863	1.839	1.818	1.799	1.781	1.766	1.751
60	1.952	1.917	1.887	1.860	1.836	1.815	1.796	1.778	1.763	1.748
61	1.949	1.915	1.884	1.857	1.834	1.812	1.793	1.776	1.760	1.745
62	1.947	1.912	1.882	1.855	1.831	1.809	1.790	1.773	1.757	1.742

TABLE B.3 (continued)
Upper Critical Values of the F Distribution for ν_1 Numerator Degrees of Freedom and ν_2 Denominator Degrees of Freedom

5% Significance Level
$F_{.05}(\nu_1, \nu_2)$

ν_1

ν_2	11	12	13	14	15	16	17	18	19	20
63	1.944	1.909	1.879	1.852	1.828	1.807	1.787	1.770	1.754	1.739
64	1.942	1.907	1.876	1.849	1.826	1.804	1.785	1.767	1.751	1.737
65	1.939	1.904	1.874	1.847	1.823	1.802	1.782	1.765	1.749	1.734
66	1.937	1.902	1.871	1.845	1.821	1.799	1.780	1.762	1.746	1.732
67	1.935	1.900	1.869	1.842	1.818	1.797	1.777	1.760	1.744	1.729
68	1.932	1.897	1.867	1.840	1.816	1.795	1.775	1.758	1.742	1.727
69	1.930	1.895	1.865	1.838	1.814	1.792	1.773	1.755	1.739	1.725
70	1.928	1.893	1.863	1.836	1.812	1.790	1.771	1.753	1.737	1.722
71	1.926	1.891	1.861	1.834	1.810	1.788	1.769	1.751	1.735	1.720
72	1.924	1.889	1.859	1.832	1.808	1.786	1.767	1.749	1.733	1.718
73	1.922	1.887	1.857	1.830	1.806	1.784	1.765	1.747	1.731	1.716
74	1.921	1.885	1.855	1.828	1.804	1.782	1.763	1.745	1.729	1.714
75	1.919	1.884	1.853	1.826	1.802	1.780	1.761	1.743	1.727	1.712
76	1.917	1.882	1.851	1.824	1.800	1.778	1.759	1.741	1.725	1.710
77	1.915	1.880	1.849	1.822	1.798	1.777	1.757	1.739	1.723	1.708
78	1.914	1.878	1.848	1.821	1.797	1.775	1.755	1.738	1.721	1.707
79	1.912	1.877	1.846	1.819	1.795	1.773	1.754	1.736	1.720	1.705
80	1.910	1.875	1.845	1.817	1.793	1.772	1.752	1.734	1.718	1.703
81	1.909	1.874	1.843	1.815	1.792	1.770	1.750	1.733	1.716	1.702

df										
82	1.700	1.715	1.731	1.749	1.768	1.790	1.814	1.841	1.872	1.907
83	1.698	1.713	1.729	1.747	1.767	1.789	1.813	1.840	1.871	1.906
84	1.697	1.712	1.728	1.746	1.765	1.787	1.811	1.838	1.869	1.905
85	1.695	1.710	1.726	1.744	1.764	1.786	1.810	1.837	1.868	1.903
86	1.694	1.709	1.725	1.743	1.762	1.784	1.808	1.836	1.867	1.902
87	1.692	1.707	1.724	1.741	1.761	1.783	1.807	1.834	1.865	1.900
88	1.691	1.706	1.722	1.740	1.760	1.782	1.806	1.833	1.864	1.899
89	1.690	1.705	1.721	1.739	1.758	1.780	1.804	1.832	1.863	1.898
90	1.688	1.703	1.720	1.737	1.757	1.779	1.803	1.830	1.861	1.897
91	1.687	1.702	1.718	1.736	1.756	1.778	1.802	1.829	1.860	1.895
92	1.686	1.701	1.717	1.735	1.755	1.776	1.801	1.828	1.859	1.894
93	1.684	1.699	1.716	1.734	1.753	1.775	1.800	1.827	1.858	1.893
94	1.683	1.698	1.715	1.733	1.752	1.774	1.798	1.826	1.857	1.892
95	1.682	1.697	1.713	1.731	1.751	1.773	1.797	1.825	1.856	1.891
96	1.681	1.696	1.712	1.730	1.750	1.772	1.796	1.823	1.854	1.890
97	1.680	1.695	1.711	1.729	1.749	1.771	1.795	1.822	1.853	1.889
98	1.679	1.694	1.710	1.728	1.748	1.770	1.794	1.821	1.852	1.888
99	1.678	1.693	1.709	1.727	1.747	1.769	1.793	1.820	1.851	1.887
100	1.676	1.691	1.708	1.726	1.746	1.768	1.792	1.819	1.850	1.886

TABLE B.3 (continued)
Upper Critical Values of the F Distribution for v_1 Numerator Degrees of Freedom and v_2 Denominator Degrees of Freedom

1% Significance Level
$F_{.01}(v_1, v_2)$

v_2	v_1 1	2	3	4	5	6	7	8	9	10
1	4052.19	4999.52	5403.34	5624.62	5763.65	5858.97	5928.33	5981.10	6022.50	6055.85
2	98.502	99.000	99.166	99.249	99.300	99.333	99.356	99.374	99.388	99.399
3	34.116	30.816	29.457	28.710	28.237	27.911	27.672	27.489	27.345	27.229
4	21.198	18.000	16.694	15.977	15.522	15.207	14.976	14.799	14.659	14.546
5	16.258	13.274	12.060	11.392	10.967	10.672	10.456	10.289	10.158	10.051
6	13.745	10.925	9.780	9.148	8.746	8.466	8.260	8.102	7.976	7.874
7	12.246	9.547	8.451	7.847	7.460	7.191	6.993	6.840	6.719	6.620
8	11.259	8.649	7.591	7.006	6.632	6.371	6.178	6.029	5.911	5.814
9	10.561	8.022	6.992	6.422	6.057	5.802	5.613	5.467	5.351	5.257
10	10.044	7.559	6.552	5.994	5.636	5.386	5.200	5.057	4.942	4.849
11	9.646	7.206	6.217	5.668	5.316	5.069	4.886	4.744	4.632	4.539
12	9.330	6.927	5.953	5.412	5.064	4.821	4.640	4.499	4.388	4.296
13	9.074	6.701	5.739	5.205	4.862	4.620	4.441	4.302	4.191	4.100
14	8.862	6.515	5.564	5.035	4.695	4.456	4.278	4.140	4.030	3.939
15	8.683	6.359	5.417	4.893	4.556	4.318	4.142	4.004	3.895	3.805
16	8.531	6.226	5.292	4.773	4.437	4.202	4.026	3.890	3.780	3.691
17	8.400	6.112	5.185	4.669	4.336	4.102	3.927	3.791	3.682	3.593
18	8.285	6.013	5.092	4.579	4.248	4.015	3.841	3.705	3.597	3.508
19	8.185	5.926	5.010	4.500	4.171	3.939	3.765	3.631	3.523	3.434

20	8.096	5.849	4.938	4.431	4.103	3.871	3.699	3.564	3.457	3.368
21	8.017	5.780	4.874	4.369	4.042	3.812	3.640	3.506	3.398	3.310
22	7.945	5.719	4.817	4.313	3.988	3.758	3.587	3.453	3.346	3.258
23	7.881	5.664	4.765	4.264	3.939	3.710	3.539	3.406	3.299	3.211
24	7.823	5.614	4.718	4.218	3.895	3.667	3.496	3.363	3.256	3.168
25	7.770	5.568	4.675	4.177	3.855	3.627	3.457	3.324	3.217	3.129
26	7.721	5.526	4.637	4.140	3.818	3.591	3.421	3.288	3.182	3.094
27	7.677	5.488	4.601	4.106	3.785	3.558	3.388	3.256	3.149	3.062
28	7.636	5.453	4.568	4.074	3.754	3.528	3.358	3.226	3.120	3.032
29	7.598	5.420	4.538	4.045	3.725	3.499	3.330	3.198	3.092	3.005
30	7.562	5.390	4.510	4.018	3.699	3.473	3.305	3.173	3.067	2.979
31	7.530	5.362	4.484	3.993	3.675	3.449	3.281	3.149	3.043	2.955
32	7.499	5.336	4.459	3.969	3.652	3.427	3.258	3.127	3.021	2.934
33	7.471	5.312	4.437	3.948	3.630	3.406	3.238	3.106	3.000	2.913
34	7.444	5.289	4.416	3.927	3.611	3.386	3.218	3.087	2.981	2.894
35	7.419	5.268	4.396	3.908	3.592	3.368	3.200	3.069	2.963	2.876
36	7.396	5.248	4.377	3.890	3.574	3.351	3.183	3.052	2.946	2.859
37	7.373	5.229	4.360	3.873	3.558	3.334	3.167	3.036	2.930	2.843
38	7.353	5.211	4.343	3.858	3.542	3.319	3.152	3.021	2.915	2.828
39	7.333	5.194	4.327	3.843	3.528	3.305	3.137	3.006	2.901	2.814
40	7.314	5.179	4.313	3.828	3.514	3.291	3.124	2.993	2.888	2.801
41	7.296	5.163	4.299	3.815	3.501	3.278	3.111	2.980	2.875	2.788
42	7.280	5.149	4.285	3.802	3.488	3.266	3.099	2.968	2.863	2.776
43	7.264	5.136	4.273	3.790	3.476	3.254	3.087	2.957	2.851	2.764
44	7.248	5.123	4.261	3.778	3.465	3.243	3.076	2.946	2.840	2.754
45	7.234	5.110	4.249	3.767	3.454	3.232	3.066	2.935	2.830	2.743
46	7.220	5.099	4.238	3.757	3.444	3.222	3.056	2.925	2.820	2.733
47	7.207	5.087	4.228	3.747	3.434	3.213	3.046	2.916	2.811	2.724
48	7.194	5.077	4.218	3.737	3.425	3.204	3.037	2.907	2.802	2.715
49	7.182	5.066	4.208	3.728	3.416	3.195	3.028	2.898	2.793	2.706

TABLE B.3 (continued)
Upper Critical Values of the F Distribution for v_1 Numerator Degrees of Freedom and v_2 Denominator Degrees of Freedom

1% Significance Level
$F_{.01}(v_1, v_2)$

v_2	v_1									
	1	2	3	4	5	6	7	8	9	10
50	7.171	5.057	4.199	3.720	3.408	3.186	3.020	2.890	2.785	2.698
51	7.159	5.047	4.191	3.711	3.400	3.178	3.012	2.882	2.777	2.690
52	7.149	5.038	4.182	3.703	3.392	3.171	3.005	2.874	2.769	2.683
53	7.139	5.030	4.174	3.695	3.384	3.163	2.997	2.867	2.762	2.675
54	7.129	5.021	4.167	3.688	3.377	3.156	2.990	2.860	2.755	2.668
55	7.119	5.013	4.159	3.681	3.370	3.149	2.983	2.853	2.748	2.662
56	7.110	5.006	4.152	3.674	3.363	3.143	2.977	2.847	2.742	2.655
57	7.102	4.998	4.145	3.667	3.357	3.136	2.971	2.841	2.736	2.649
58	7.093	4.991	4.138	3.661	3.351	3.130	2.965	2.835	2.730	2.643
59	7.085	4.984	4.132	3.655	3.345	3.124	2.959	2.829	2.724	2.637
60	7.077	4.977	4.126	3.649	3.339	3.119	2.953	2.823	2.718	2.632
61	7.070	4.971	4.120	3.643	3.333	3.113	2.948	2.818	2.713	2.626
62	7.062	4.965	4.114	3.638	3.328	3.108	2.942	2.813	2.708	2.621
63	7.055	4.959	4.109	3.632	3.323	3.103	2.937	2.808	2.703	2.616
64	7.048	4.953	4.103	3.627	3.318	3.098	2.932	2.803	2.698	2.611
65	7.042	4.947	4.098	3.622	3.313	3.093	2.928	2.798	2.693	2.607
66	7.035	4.942	4.093	3.618	3.308	3.088	2.923	2.793	2.689	2.602
67	7.029	4.937	4.088	3.613	3.304	3.084	2.919	2.789	2.684	2.598
68	7.023	4.932	4.083	3.608	3.299	3.080	2.914	2.785	2.680	2.593
69	7.017	4.927	4.079	3.604	3.295	3.075	2.910	2.781	2.676	2.589

70	2.585	2.672	2.777	2.906	3.071	3.291	3.600	4.074	4.922	7.011
71	2.581	2.668	2.773	2.902	3.067	3.287	3.596	4.070	4.917	7.006
72	2.578	2.664	2.769	2.898	3.063	3.283	3.591	4.066	4.913	7.001
73	2.574	2.660	2.765	2.895	3.060	3.279	3.588	4.062	4.908	6.995
74	2.570	2.657	2.762	2.891	3.056	3.275	3.584	4.058	4.904	6.990
75	2.567	2.653	2.758	2.887	3.052	3.272	3.580	4.054	4.900	6.985
76	2.563	2.650	2.755	2.884	3.049	3.268	3.577	4.050	4.896	6.981
77	2.560	2.647	2.751	2.881	3.046	3.265	3.573	4.047	4.892	6.976
78	2.557	2.644	2.748	2.877	3.042	3.261	3.570	4.043	4.888	6.971
79	2.554	2.640	2.745	2.874	3.039	3.258	3.566	4.040	4.884	6.967
80	2.551	2.637	2.742	2.871	3.036	3.255	3.563	4.036	4.881	6.963
81	2.548	2.634	2.739	2.868	3.033	3.252	3.560	4.033	4.877	6.958
82	2.545	2.632	2.736	2.865	3.030	3.249	3.557	4.030	4.874	6.954
83	2.542	2.629	2.733	2.863	3.027	3.246	3.554	4.027	4.870	6.950
84	2.539	2.626	2.731	2.860	3.025	3.243	3.551	4.024	4.867	6.947
85	2.537	2.623	2.728	2.857	3.022	3.240	3.548	4.021	4.864	6.943
86	2.534	2.621	2.725	2.854	3.019	3.238	3.545	4.018	4.861	6.939
87	2.532	2.618	2.723	2.852	3.017	3.235	3.543	4.015	4.858	6.935
88	2.529	2.616	2.720	2.849	3.014	3.233	3.540	4.012	4.855	6.932
89	2.527	2.613	2.718	2.847	3.012	3.230	3.538	4.010	4.852	6.928
90	2.524	2.611	2.715	2.845	3.009	3.228	3.535	4.007	4.849	6.925
91	2.522	2.609	2.713	2.842	3.007	3.225	3.533	4.004	4.846	6.922
92	2.520	2.606	2.711	2.840	3.004	3.223	3.530	4.002	4.844	6.919
93	2.518	2.604	2.709	2.838	3.002	3.221	3.528	3.999	4.841	6.915
94	2.515	2.602	2.706	2.835	3.000	3.218	3.525	3.997	4.838	6.912
95	2.513	2.600	2.704	2.833	2.998	3.216	3.523	3.995	4.836	6.909
96	2.511	2.598	2.702	2.831	2.996	3.214	3.521	3.992	4.833	6.906
97	2.509	2.596	2.700	2.829	2.994	3.212	3.519	3.990	4.831	6.904
98	2.507	2.594	2.698	2.827	2.992	3.210	3.517	3.988	4.829	6.901
99	2.505	2.592	2.696	2.825	2.990	3.208	3.515	3.986	4.826	6.898
100	2.503	2.590	2.694	2.823	2.988	3.206	3.513	3.984	4.824	6.895

TABLE B.3 (continued)
Upper Critical Values of the F Distribution for v_1 Numerator Degrees of Freedom and v_2 Denominator Degrees of Freedom

1% Significance Level
$$F_{.01}(v_1, v_2)$$

v_1

v_2	11	12	13	14	15	16	17	18	19	20
1	6083.35	6106.35	6125.86	6142.70	6157.28	6170.12	6181.42	6191.52	6200.58	6208.74
2	99.408	99.416	99.422	99.428	99.432	99.437	99.440	99.444	99.447	99.449
3	27.133	27.052	26.983	26.924	26.872	26.827	26.787	26.751	26.719	26.690
4	14.452	14.374	14.307	14.249	14.198	14.154	14.115	14.080	14.048	14.020
5	9.963	9.888	9.825	9.770	9.722	9.680	9.643	9.610	9.580	9.553
6	7.790	7.718	7.657	7.605	7.559	7.519	7.483	7.451	7.422	7.396
7	6.538	6.469	6.410	6.359	6.314	6.275	6.240	6.209	6.181	6.155
8	5.734	5.667	5.609	5.559	5.515	5.477	5.442	5.412	5.384	5.359
9	5.178	5.111	5.055	5.005	4.962	4.924	4.890	4.860	4.833	4.808
10	4.772	4.706	4.650	4.601	4.558	4.520	4.487	4.457	4.430	4.405
11	4.462	4.397	4.342	4.293	4.251	4.213	4.180	4.150	4.123	4.099
12	4.220	4.155	4.100	4.052	4.010	3.972	3.939	3.909	3.883	3.858
13	4.025	3.960	3.905	3.857	3.815	3.778	3.745	3.716	3.689	3.665
14	3.864	3.800	3.745	3.698	3.656	3.619	3.586	3.556	3.529	3.505
15	3.730	3.666	3.612	3.564	3.522	3.485	3.452	3.423	3.396	3.372
16	3.616	3.553	3.498	3.451	3.409	3.372	3.339	3.310	3.283	3.259
17	3.519	3.455	3.401	3.353	3.312	3.275	3.242	3.212	3.186	3.162
18	3.434	3.371	3.316	3.269	3.227	3.190	3.158	3.128	3.101	3.077
19	3.360	3.297	3.242	3.195	3.153	3.116	3.084	3.054	3.027	3.003
20	3.294	3.231	3.177	3.130	3.088	3.051	3.018	2.989	2.962	2.938

21	3.236	3.173	3.119	3.072	3.030	2.993	2.960	2.931	2.904	2.880
22	3.184	3.121	3.067	3.019	2.978	2.941	2.908	2.879	2.852	2.827
23	3.137	3.074	3.020	2.973	2.931	2.894	2.861	2.832	2.805	2.781
24	3.094	3.032	2.977	2.930	2.889	2.852	2.819	2.789	2.762	2.738
25	3.056	2.993	2.939	2.892	2.850	2.813	2.780	2.751	2.724	2.699
26	3.021	2.958	2.904	2.857	2.815	2.778	2.745	2.715	2.688	2.664
27	2.988	2.926	2.871	2.824	2.783	2.746	2.713	2.683	2.656	2.632
28	2.959	2.896	2.842	2.795	2.753	2.716	2.683	2.653	2.626	2.602
29	2.931	2.868	2.814	2.767	2.726	2.689	2.656	2.626	2.599	2.574
30	2.906	2.843	2.789	2.742	2.700	2.663	2.630	2.600	2.573	2.549
31	2.882	2.820	2.765	2.718	2.677	2.640	2.606	2.577	2.550	2.525
32	2.860	2.798	2.744	2.696	2.655	2.618	2.584	2.555	2.527	2.503
33	2.840	2.777	2.723	2.676	2.634	2.597	2.564	2.534	2.507	2.482
34	2.821	2.758	2.704	2.657	2.615	2.578	2.545	2.515	2.488	2.463
35	2.803	2.740	2.686	2.639	2.597	2.560	2.527	2.497	2.470	2.445
36	2.786	2.723	2.669	2.622	2.580	2.543	2.510	2.480	2.453	2.428
37	2.770	2.707	2.653	2.606	2.564	2.527	2.494	2.464	2.437	2.412
38	2.755	2.692	2.638	2.591	2.549	2.512	2.479	2.449	2.421	2.397
39	2.741	2.678	2.624	2.577	2.535	2.498	2.465	2.434	2.407	2.382
40	2.727	2.665	2.611	2.563	2.522	2.484	2.451	2.421	2.394	2.369
41	2.715	2.652	2.598	2.551	2.509	2.472	2.438	2.408	2.381	2.356
42	2.703	2.640	2.586	2.539	2.497	2.460	2.426	2.396	2.369	2.344
43	2.691	2.629	2.575	2.527	2.485	2.448	2.415	2.385	2.357	2.332
44	2.680	2.618	2.564	2.516	2.475	2.437	2.404	2.374	2.346	2.321
45	2.670	2.608	2.553	2.506	2.464	2.427	2.393	2.363	2.336	2.311
46	2.660	2.598	2.544	2.496	2.454	2.417	2.384	2.353	2.326	2.301
47	2.651	2.588	2.534	2.487	2.445	2.408	2.374	2.344	2.316	2.291
48	2.642	2.579	2.525	2.478	2.436	2.399	2.365	2.335	2.307	2.282
49	2.633	2.571	2.517	2.469	2.427	2.390	2.356	2.326	2.299	2.274
50	2.625	2.562	2.508	2.461	2.419	2.382	2.348	2.318	2.290	2.265

TABLE B.3 (continued)
Upper Critical Values of the F Distribution for v_1 Numerator Degrees of Freedom and v_2 Denominator Degrees of Freedom

1% Significance Level
$F_{.01}(v_1, v_2)$

v_2	v_1 11	12	13	14	15	16	17	18	19	20
51	2.617	2.555	2.500	2.453	2.411	2.374	2.340	2.310	2.282	2.257
52	2.610	2.547	2.493	2.445	2.403	2.366	2.333	2.302	2.275	2.250
53	2.602	2.540	2.486	2.438	2.396	2.359	2.325	2.295	2.267	2.242
54	2.595	2.533	2.479	2.431	2.389	2.352	2.318	2.288	2.260	2.235
55	2.589	2.526	2.472	2.424	2.382	2.345	2.311	2.281	2.253	2.228
56	2.582	2.520	2.465	2.418	2.376	2.339	2.305	2.275	2.247	2.222
57	2.576	2.513	2.459	2.412	2.370	2.332	2.299	2.268	2.241	2.215
58	2.570	2.507	2.453	2.406	2.364	2.326	2.293	2.262	2.235	2.209
59	2.564	2.502	2.447	2.400	2.358	2.320	2.287	2.256	2.229	2.203
60	2.559	2.496	2.442	2.394	2.352	2.315	2.281	2.251	2.223	2.198
61	2.553	2.491	2.436	2.389	2.347	2.309	2.276	2.245	2.218	2.192
62	2.548	2.486	2.431	2.384	2.342	2.304	2.270	2.240	2.212	2.187
63	2.543	2.481	2.426	2.379	2.337	2.299	2.265	2.235	2.207	2.182
64	2.538	2.476	2.421	2.374	2.332	2.294	2.260	2.230	2.202	2.177
65	2.534	2.471	2.417	2.369	2.327	2.289	2.256	2.225	2.198	2.172
66	2.529	2.466	2.412	2.365	2.322	2.285	2.251	2.221	2.193	2.168
67	2.525	2.462	2.408	2.360	2.318	2.280	2.247	2.216	2.188	2.163
68	2.520	2.458	2.403	2.356	2.314	2.276	2.242	2.212	2.184	2.159
69	2.516	2.454	2.399	2.352	2.310	2.272	2.238	2.208	2.180	2.155
70	2.512	2.450	2.395	2.348	2.306	2.268	2.234	2.204	2.176	2.150
71	2.508	2.446	2.391	2.344	2.302	2.264	2.230	2.200	2.172	2.146

72	2.504	2.442	2.388	2.340	2.298	2.260	2.226	2.196	2.168	2.143
73	2.501	2.438	2.384	2.336	2.294	2.256	2.223	2.192	2.164	2.139
74	2.497	2.435	2.380	2.333	2.290	2.253	2.219	2.188	2.161	2.135
75	2.494	2.431	2.377	2.329	2.287	2.249	2.215	2.185	2.157	2.132
76	2.490	2.428	2.373	2.326	2.284	2.246	2.212	2.181	2.154	2.128
77	2.487	2.424	2.370	2.322	2.280	2.243	2.209	2.178	2.150	2.125
78	2.484	2.421	2.367	2.319	2.277	2.239	2.206	2.175	2.147	2.122
79	2.481	2.418	2.364	2.316	2.274	2.236	2.202	2.172	2.144	2.118
80	2.478	2.415	2.361	2.313	2.271	2.233	2.199	2.169	2.141	2.115
81	2.475	2.412	2.358	2.310	2.268	2.230	2.196	2.166	2.138	2.112
82	2.472	2.409	2.355	2.307	2.265	2.227	2.193	2.163	2.135	2.109
83	2.469	2.406	2.352	2.304	2.262	2.224	2.191	2.160	2.132	2.106
84	2.466	2.404	2.349	2.302	2.259	2.222	2.188	2.157	2.129	2.104
85	2.464	2.401	2.347	2.299	2.257	2.219	2.185	2.154	2.126	2.101
86	2.461	2.398	2.344	2.296	2.254	2.216	2.182	2.152	2.124	2.098
87	2.459	2.396	2.342	2.294	2.252	2.214	2.180	2.149	2.121	2.096
88	2.456	2.393	2.339	2.291	2.249	2.211	2.177	2.147	2.119	2.093
89	2.454	2.391	2.337	2.289	2.247	2.209	2.175	2.144	2.116	2.091
90	2.451	2.389	2.334	2.286	2.244	2.206	2.172	2.142	2.114	2.088
91	2.449	2.386	2.332	2.284	2.242	2.204	2.170	2.139	2.111	2.086
92	2.447	2.384	2.330	2.282	2.240	2.202	2.168	2.137	2.109	2.083
93	2.444	2.382	2.327	2.280	2.237	2.200	2.166	2.135	2.107	2.081
94	2.442	2.380	2.325	2.277	2.235	2.197	2.163	2.133	2.105	2.079
95	2.440	2.378	2.323	2.275	2.233	2.195	2.161	2.130	2.102	2.077
96	2.438	2.375	2.321	2.273	2.231	2.193	2.159	2.128	2.100	2.075
97	2.436	2.373	2.319	2.271	2.229	2.191	2.157	2.126	2.098	2.073
98	2.434	2.371	2.317	2.269	2.227	2.189	2.155	2.124	2.096	2.071
99	2.432	2.369	2.315	2.267	2.225	2.187	2.153	2.122	2.094	2.069
100	2.430	2.368	2.313	2.265	2.223	2.185	2.151	2.120	2.092	2.067

Adapted from the *National Institute of Standards and Technology Engineering Statistics Handbook.*

B.4 CRITICAL VALUES OF THE CHI-SQUARE DISTRIBUTION

B.4.1 How to Use This Table

Table B.4 contains the critical values of the chi-square distribution. Because of the lack of symmetry of the chi-square distribution, the table is divided into two parts: for the upper and lower tails of the distribution.

A test statistic with v degrees of freedom is computed from the data. For upper one-sided tests, the test statistic is compared with a value from the upper critical values portion of the table. For two-sided tests, the test statistic is compared with values from both parts of the table.

The significance level, α, is demonstrated with the graph below, which shows a chi-square distribution with 3 degrees of freedom for a two-sided test at significance level $\alpha = .05$. If the test statistic is greater than the upper critical value or less than the lower critical value, we reject the null hypothesis. Specific instructions are given below.

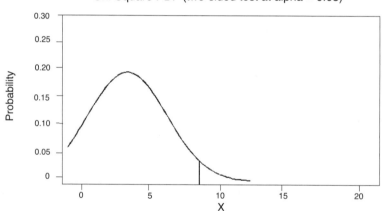

Chi-Square PDF (two-sided test at alpha = 0.05)

Given a specified value for α:

For a two-sided test, find the column corresponding to $\alpha/2$ in the upper critical values section of the table and reject the null hypothesis if the test statistic is greater than the tabled value. Similarly, find the column corresponding to $1 - \alpha/2$ in the lower critical values portion of the table and reject the null hypothesis if the test statistic is less than the tabled value.

For an upper one-sided test, find the column corresponding to α in the upper critical values section and reject the null hypothesis if the test statistic is greater than the tabled value.

For a lower one-sided test, find the column corresponding to $1 - \alpha$ in the lower critical values section and reject the null hypothesis if the computed test statistic is less than the tabled value.

TABLE B.4
Critical Values of Chi-Square Distribution with
v Degrees of Freedom

Upper Critical Values

Probability of exceeding the critical value

v	0.10	0.05	0.025	0.01	0.001
1	2.706	3.841	5.024	6.635	10.828
2	4.605	5.991	7.378	9.210	13.816
3	6.251	7.815	9.348	11.345	16.266
4	7.779	9.488	11.143	13.277	18.467
5	9.236	11.070	12.833	15.086	20.515
6	10.645	12.592	14.449	16.812	22.458
7	12.017	14.067	16.013	18.475	24.322
8	13.362	15.507	17.535	20.090	26.125
9	14.684	16.919	19.023	21.666	27.877
10	15.987	18.307	20.483	23.209	29.588
11	17.275	19.675	21.920	24.725	31.264
12	18.549	21.026	23.337	26.217	32.910
13	19.812	22.362	24.736	27.688	34.528
14	21.064	23.685	26.119	29.141	36.123
15	22.307	24.996	27.488	30.578	37.697
16	23.542	26.296	28.845	32.000	39.252
17	24.769	27.587	30.191	33.409	40.790
18	25.989	28.869	31.526	34.805	42.312
19	27.204	30.144	32.852	36.191	43.820
20	28.412	31.410	34.170	37.566	45.315
21	29.615	32.671	35.479	38.932	46.797
22	30.813	33.924	36.781	40.289	48.268
23	32.007	35.172	38.076	41.638	49.728
24	33.196	36.415	39.364	42.980	51.179
25	34.382	37.652	40.646	44.314	52.620
26	35.563	38.885	41.923	45.642	54.052
27	36.741	40.113	43.195	46.963	55.476
28	37.916	41.337	44.461	48.278	56.892
29	39.087	42.557	45.722	49.588	58.301
30	40.256	43.773	46.979	50.892	59.703
31	41.422	44.985	48.232	52.191	61.098
32	42.585	46.194	49.480	53.486	62.487
33	43.745	47.400	50.725	54.776	63.870
34	44.903	48.602	51.966	56.061	65.247
35	46.059	49.802	53.203	57.342	66.619
36	47.212	50.998	54.437	58.619	67.985
37	48.363	52.192	55.668	59.893	69.347
38	49.513	53.384	56.896	61.162	70.703
39	50.660	54.572	58.120	62.428	72.055
40	51.805	55.758	59.342	63.691	73.402

TABLE B.4 (continued)
Critical Values of Chi-Square Distribution with
v Degrees of Freedom

Upper Critical Values

Probability of exceeding the critical value

v	0.10	0.05	0.025	0.01	0.001
41	52.949	56.942	60.561	64.950	74.745
42	54.090	58.124	61.777	66.206	76.084
43	55.230	59.304	62.990	67.459	77.419
44	56.369	60.481	64.201	68.710	78.750
45	57.505	61.656	65.410	69.957	80.077
46	58.641	62.830	66.617	71.201	81.400
47	59.774	64.001	67.821	72.443	82.720
48	60.907	65.171	69.023	73.683	84.037
49	62.038	66.339	70.222	74.919	85.351
50	63.167	67.505	71.420	76.154	86.661
51	64.295	68.669	72.616	77.386	87.968
52	65.422	69.832	73.810	78.616	89.272
53	66.548	70.993	75.002	79.843	90.573
54	67.673	72.153	76.192	81.069	91.872
55	68.796	73.311	77.380	82.292	93.168
56	69.919	74.468	78.567	83.513	94.461
57	71.040	75.624	79.752	84.733	95.751
58	72.160	76.778	80.936	85.950	97.039
59	73.279	77.931	82.117	87.166	98.324
60	74.397	79.082	83.298	88.379	99.607
61	75.514	80.232	84.476	89.591	100.888
62	76.630	81.381	85.654	90.802	102.166
63	77.745	82.529	86.830	92.010	103.442
64	78.860	83.675	88.004	93.217	104.716
65	79.973	84.821	89.177	94.422	105.988
66	81.085	85.965	90.349	95.626	107.258
67	82.197	87.108	91.519	96.828	108.526
68	83.308	88.250	92.689	98.028	109.791
69	84.418	89.391	93.856	99.228	111.055
70	85.527	90.531	95.023	100.425	112.317
71	86.635	91.670	96.189	101.621	113.577
72	87.743	92.808	97.353	102.816	114.835
73	88.850	93.945	98.516	104.010	116.092
74	89.956	95.081	99.678	105.202	117.346
75	91.061	96.217	100.839	106.393	118.599
76	92.166	97.351	101.999	107.583	119.850
77	93.270	98.484	103.158	108.771	121.100
78	94.374	99.617	104.316	109.958	122.348
79	95.476	100.749	105.473	111.144	123.594
80	96.578	101.879	106.629	112.329	124.839

TABLE B.4 (continued)
Critical Values of Chi-Square Distribution with
v Degrees of Freedom

Upper Critical Values

Probability of exceeding the critical value

v	0.10	0.05	0.025	0.01	0.001
81	97.680	103.010	107.783	113.512	126.083
82	98.780	104.139	108.937	114.695	127.324
83	99.880	105.267	110.090	115.876	128.565
84	100.980	106.395	111.242	117.057	129.804
85	102.079	107.522	112.393	118.236	131.041
86	103.177	108.648	113.544	119.414	132.277
87	104.275	109.773	114.693	120.591	133.512
88	105.372	110.898	115.841	121.767	134.746
89	106.469	112.022	116.989	122.942	135.978
90	107.565	113.145	118.136	124.116	137.208
91	108.661	114.268	119.282	125.289	138.438
92	109.756	115.390	120.427	126.462	139.666
93	110.850	116.511	121.571	127.633	140.893
94	111.944	117.632	122.715	128.803	142.119
95	113.038	118.752	123.858	129.973	143.344
96	114.131	119.871	125.000	131.141	144.567
97	115.223	120.990	126.141	132.309	145.789
98	116.315	122.108	127.282	133.476	147.010
99	117.407	123.225	128.422	134.642	148.230
100	118.498	124.342	129.561	135.807	149.449

Lower Critical Values

Probability of exceeding the critical value

v	0.90	0.95	0.975	0.99	0.999
1	.016	.004	.001	.000	.000
2	.211	.103	.051	.020	.002
3	.584	.352	.216	.115	.024
4	1.064	.711	.484	.297	.091
5	1.610	1.145	.831	.554	.210
6	2.204	1.635	1.237	.872	.381
7	2.833	2.167	1.690	1.239	.598
8	3.490	2.733	2.180	1.646	.857
9	4.168	3.325	2.700	2.088	1.152
10	4.865	3.940	3.247	2.558	1.479
11	5.578	4.575	3.816	3.053	1.834
12	6.304	5.226	4.404	3.571	2.214
13	7.042	5.892	5.009	4.107	2.617
14	7.790	6.571	5.629	4.660	3.041

TABLE B.4 (continued)
Critical Values of Chi-Square Distribution with
ν Degrees of Freedom

Lower Critical Values

Probability of exceeding the critical value

ν	0.90	0.95	0.975	0.99	0.999
15	8.547	7.261	6.262	5.229	3.483
16	9.312	7.962	6.908	5.812	3.942
17	10.085	8.672	7.564	6.408	4.416
18	10.865	9.390	8.231	7.015	4.905
19	11.651	10.117	8.907	7.633	5.407
20	12.443	10.851	9.591	8.260	5.921
21	13.240	11.591	10.283	8.897	6.447
22	14.041	12.338	10.982	9.542	6.983
23	14.848	13.091	11.689	10.196	7.529
24	15.659	13.848	12.401	10.856	8.085
25	16.473	14.611	13.120	11.524	8.649
26	17.292	15.379	13.844	12.198	9.222
27	18.114	16.151	14.573	12.879	9.803
28	18.939	16.928	15.308	13.565	10.391
29	19.768	17.708	16.047	14.256	10.986
30	20.599	18.493	16.791	14.953	11.588
31	21.434	19.281	17.539	15.655	12.196
32	22.271	20.072	18.291	16.362	12.811
33	23.110	20.867	19.047	17.074	13.431
34	23.952	21.664	19.806	17.789	14.057
35	24.797	22.465	20.569	18.509	14.688
36	25.643	23.269	21.336	19.233	15.324
37	26.492	24.075	22.106	19.960	15.965
38	27.343	24.884	22.878	20.691	16.611
39	28.196	25.695	23.654	21.426	17.262
40	29.051	26.509	24.433	22.164	17.916
41	29.907	27.326	25.215	22.906	18.575
42	30.765	28.144	25.999	23.650	19.239
43	31.625	28.965	26.785	24.398	19.906
44	32.487	29.787	27.575	25.148	20.576
45	33.350	30.612	28.366	25.901	21.251
46	34.215	31.439	29.160	26.657	21.929
47	35.081	32.268	29.956	27.416	22.610
48	35.949	33.098	30.755	28.177	23.295
49	36.818	33.930	31.555	28.941	23.983
50	37.689	34.764	32.357	29.707	24.674
51	38.560	35.600	33.162	30.475	25.368
52	39.433	36.437	33.968	31.246	26.065
53	40.308	37.276	34.776	32.018	26.765
54	41.183	38.116	35.586	32.793	27.468

TABLE B.4 (continued)
Critical Values of Chi-Square Distribution with
v Degrees of Freedom

Lower Critical Values

Probability of exceeding the critical value

v	0.90	0.95	0.975	0.99	0.999
55	42.060	38.958	36.398	33.570	28.173
56	42.937	39.801	37.212	34.350	28.881
57	43.816	40.646	38.027	35.131	29.592
58	44.696	41.492	38.844	35.913	30.305
59	45.577	42.339	39.662	36.698	31.020
60	46.459	43.188	40.482	37.485	31.738
61	47.342	44.038	41.303	38.273	32.459
62	48.226	44.889	42.126	39.063	33.181
63	49.111	45.741	42.950	39.855	33.906
64	49.996	46.595	43.776	40.649	34.633
65	50.883	47.450	44.603	41.444	35.362
66	51.770	48.305	45.431	42.240	36.093
67	52.659	49.162	46.261	43.038	36.826
68	53.548	50.020	47.092	43.838	37.561
69	54.438	50.879	47.924	44.639	38.298
70	55.329	51.739	48.758	45.442	39.036
71	56.221	52.600	49.592	46.246	39.777
72	57.113	53.462	50.428	47.051	40.519
73	58.006	54.325	51.265	47.858	41.264
74	58.900	55.189	52.103	48.666	42.010
75	59.795	56.054	52.942	49.475	42.757
76	60.690	56.920	53.782	50.286	43.507
77	61.586	57.786	54.623	51.097	44.258
78	62.483	58.654	55.466	51.910	45.010
79	63.380	59.522	56.309	52.725	45.764
80	64.278	60.391	57.153	53.540	46.520
81	65.176	61.261	57.998	54.357	47.277
82	66.076	62.132	58.845	55.174	48.036
83	66.976	63.004	59.692	55.993	48.796
84	67.876	63.876	60.540	56.813	49.557
85	68.777	64.749	61.389	57.634	50.320
86	69.679	65.623	62.239	58.456	51.085
87	70.581	66.498	63.089	59.279	51.850
88	71.484	67.373	63.941	60.103	52.617
89	72.387	68.249	64.793	60.928	53.386
90	73.291	69.126	65.647	61.754	54.155
91	74.196	70.003	66.501	62.581	54.926
92	75.100	70.882	67.356	63.409	55.698
93	76.006	71.760	68.211	64.238	56.472
94	76.912	72.640	69.068	65.068	57.246

TABLE B.4 (continued)
Critical Values of Chi-Square Distribution with ν Degrees of Freedom

Lower Critical Values

Probability of exceeding the critical value

ν	0.90	0.95	0.975	0.99	0.999
95	77.818	73.520	69.925	65.898	58.022
96	78.725	74.401	70.783	66.730	58.799
97	79.633	75.282	71.642	67.562	59.577
98	80.541	76.164	72.501	68.396	60.356
99	81.449	77.046	73.361	69.230	61.137
100	82.358	77.929	74.222	70.065	61.918

Adapted from the *National Institute of Standards and Technology Engineering Statistics Handbook.*

Appendix C:
TRIZ — Abbreviated Version of the 40 Principles for Inventive Problem Solving*

Principle 1. Segmentation. Fragmentation. Transition to micro level. Divide an object or system into independent parts. Make an object easy to disassemble. Increase the degree of fragmentation or segmentation.

For the last 30 years the use of teams has been one of the persistent themes in the workplace, because small teams are flexible and can make decisions quickly.

A large job can be broken down into many smaller jobs (called a "work breakdown structure" in project management). The JIT (Just-in-Time, or kanban) system uses the concept of segmentation to an extreme — it replaces the idea of mass production with the idea that the most efficient production system can produce a single unit just as easily as multiple units.

Principle 2. Separation. Separate the only necessary part (or property), or an interfering part or property, from an object or system.

Franchising separates the ownership of a local business, such as a restaurant or a printing shop, from the development of the concept and the systems that make it successful.

The ASP (application system provider) is a new concept — your company does not own its own software, but rents it as needed from a "provider." All outsourcing, including staffing services as well as production and information management services, could be considered examples of this principle.

Principle 3. Local quality. Change a system or object's structure from uniform to nonuniform; change an external environment (or external influence) from uniform to nonuniform. Make each part of an object or system function in conditions most suitable for its operation. Make each part of an object fulfill a different and useful function.

---

* Adapted from Rantanen, K., and Domb, E., *Simplified TRIZ: New Problem-Solving Applications for Engineers and Manufacturing Professionals*, St. Lucie Press, Boca Raton, FL, 2002, chap. 10. With permission.

In business, the segmentation of the market illustrates the local quality principle, too. *Segmentation* is used to divide the market into small markets with specific attributes, then *local quality* is used to treat each of those markets appropriately. The Whirlpool Company has hired marketing people who speak 18 different languages in India, to tailor its approach in the automatic washing machine market to the cultural preferences of each group.

Principle 4. Symmetry change. *Change the shape of an object or system from symmetrical to asymmetrical. If an object is asymmetrical, increase its degree of asymmetry.*

Mass customization is a business strategy that corresponds to asymmetry — the product, service, or policies of a business are specifically designed for each customer and do not need to be the same as those provided to other customers. (See articles by D. Mann and E. Domb, ETRIA, Nov. 2001 and TRIZ Journal, December 1999.)

Principle 5. Merging. *Bring closer together (or merge) identical or similar objects; assemble identical or similar parts to perform parallel operations. Make operations contiguous or parallel; bring them together in time.*

Integration in microelectronics; putting many circuit elements on the same chip.

Telephone and computer networks.

Paper sheets constitute a book, books a library.

Fragile and weak components, such as glass plates, can be made stronger without increasing weight by combining them into packages.

Principle 6. Multifunctionality or universality. *Make a part of an object or system perform multiple functions; eliminate the need for other parts. The number of parts and operations are decreased, and useful features and functions are retained.*

Asea Brown Boveri has developed an electric generator with high voltage. A conventional transformer is not needed because the generator can directly feed the electric network.

Use a single adjustable wrench for all nuts.

Principle 7. "Nested doll." *Place one object inside another; place each object, in turn, inside the other. Make one part pass through a cavity in the other. The name of this principle comes from the Russian folk art dolls, in which a series (usually six or seven) of wooden dolls are nested, one inside the other.*

Telescope structures (umbrella handles, radio antennas, pointers).

Business analogy: a boutique store inside a big market.

Principle 8. Weight compensation. *To compensate for the weight of an object or system, merge it with other objects that provide lift. To compensate for the weight of an object,* *make it interact with the environment (e.g., use aerodynamic, hydrodynamic, buoyancy, and other forces).*

Banners and signs cut so that the wind lifts them for display.

Business analogy: Compensate for the heavy organizational pyramid with project organization, process organization, temporary organization, and other less hierarchical systems "lifting" the heavy structure.

Principle 9. Preliminary counteraction. *If it will be necessary to do an action with both harmful and useful effects, this action should be replaced with anti-actions (counteractions) to control the harmful effects. Create stresses in an object or system that will oppose known undesirable working stresses later on.*

Changes and innovations usually meet resistance in the organization. Get the affected people involved so that they can participate in the planning of changes and do not feel threatened.

Principle 10. Preliminary action. *Perform, before it is needed, the required change of an object or system (either fully or partially). Pre-arrange objects such that they can come into action from the most convenient place and without losing time for their delivery.*

Preliminary perforated packaging is easy to open.

Principle 11. Beforehand compensation. *Prepare emergency means beforehand to compensate for the relatively low reliability of an object or system.*

Airbags in cars and overpressure or explosion valves in boilers.

FAQ (frequently asked questions) sections of many web sites — commonly, users are told how to help themselves to solve problems that are known to exist in the system.

Principle 12. Equipotentiality. *In a potential field, limit position changes (e.g., change operating conditions to eliminate the need to raise or lower objects in a gravity field).*

The flat factory. High shelving is not used in production.

Business analogy: a transition to a flatter organization with fewer hierarchical layers. One step in team formation is to bring all team members to the same level, to eliminate hierarchical behaviors.

Principle 13. "The other way around." *Invert the action(s) used to solve the problem (e.g., instead of cooling an object, heat it). Make movable parts (or the external environment) fixed and fixed parts movable. Turn the object or process upside-down.*

Slow food instead of fast food.

Instead of the increased traveling, working at home via the internet.

Customers find their own answers in the consultant's database, instead of having the consultant find the answer for them.

Principle 14. Curvature increase. *Instead of using rectilinear parts, surfaces, or forms, use curvilinear ones; move from flat surfaces to spherical ones; from cube or parallelepiped shapes to ball-shaped structures. Use rollers, balls, spirals, or domes. Go from linear to rotary motion. Use centrifugal forces.*

Nontechnical analogies: increasing circulation of information benefits organizational function. Curved walls and streets make neighborhoods visually identifiable (both in cities and inside large office buildings and schools).

Principle 15. Dynamic parts. *Allow (or design) the characteristics of an object, external environment, process, or system to change to be optimal or to find an optimal operating condition. Divide an object or system into parts capable of movement relative to each other. If an object (or process or system) is rigid or inflexible, make it movable or adaptive.*

In business, flexibility — the capability to make changes when the environment changes — is often the difference between success and failure. Organizations are also evolving from stiff and unchanging structures to flexible ones. Ways to increase flexibility are segmentation (Principle 1), "flatter" organizations (see Principle 12), preparing changes before encountering a problem (see Principles 9 through 11), and discarding and recovering (Principle 34).

Schools, too, have used the principle of dynamics as part of their improvement strategy. In many schools, students are no longer assigned to a fixed grade in which, for example, all 8-year-olds do third grade studies together. Rather, the curriculum is flexible. The author's nephew recently was doing fifth grade arithmetic, third grade language studies, and a personal project to learn geography, all on the same day, in a program that was based on his abilities and interests.

Principle 16. Partial or excessive actions. *If 100% of a goal is hard to achieve using a given solution method, then, by using "slightly less" or "slightly more" of the same method, the problem may be considerably easier to solve.*

If marketing cannot reach all possible customers, a solution may be to select the subgroup with the highest density of prospective buyers and concentrate efforts on them. Another solution is an excessive action: Broadcast advertising will reach people who are not potential buyers, but the target audience will be included in the group that is reached.

Principle 17. Dimensionality change. *Move an object or system in two- or three-dimensional space. Use a multi-story arrangement of objects instead of a single-story arrangement. Tilt or re-orient the object; lay it on its side. Use another side of a given area.*

Information systems may have data stored in multidimensional arrays. One, two, or three dimensions may be visible to the customer (or to the customer service employee), and the rest of the structure is hidden but helps make the data available.

Similarly, three-dimensional networks of business relationships are faster to respond to the needs of one member than one- and two-dimensional systems.

Principle 18. Mechanical vibration. *Cause an object or system to oscillate or vibrate. Increase the frequency of vibration. Use an object's resonant frequency. Use piezo-electric vibrators instead of mechanical ones. Use combined ultrasonic and electro-magnetic field oscillation.*

Technological examples: The vibration of a mobile telephone can be used instead of a ring. An object's resonant frequency is used to destroy gallstones or kidney stones by ultrasound in a technique called lithotripsy, which makes surgery unnecessary. This can also be seen as the use of segmentation, because the stone breaks itself into very small pieces, which the body then eliminates through its natural processes.

Principle 19. Periodic action. *Instead of continuous actions, use periodic or pulsating actions. If an action is already periodic, change the periodic magnitude or frequency. Use pauses between impulses to perform a different action.*

Researchers propose that taking naps in the middle of day will increase the efficiency of intellectual work.

Pauses in work can be used for training.

Principle 20. Continuity of useful action. *Carry on work continuously; make all parts of an object or system work at full load, all the time. Eliminate all idle or intermittent actions or work. Note that this principle contradicts the previous one — if you eliminate all intermittent actions, you will not have any pauses to use. This just emphasizes that the various suggestions in each principle must be applied with common sense to the particular situation.*

The changing character of manufacturing shows considerable influence of this principle. Lean and JIT Manufacturing emphasize small, customized production runs instead of long series.

Principle 21. Hurrying. Skipping. *Conduct a process or certain stages (e.g., destructive, harmful, or hazardous operations) at high speed.*

In surgery, it is well known that the longer the patient is anesthetized, the higher the risk of failure and future complications. Open-heart surgery that once took 8 h or more is now done in less than 1 h, using combinations of new tools and methods.

In business, it may sometimes be more important to act quickly than to make things slowly but with no mistakes. JR Watson, IBM founder, put it as follows: "If you want to succeed, double your failure rate." However, some companies have gone too far in introducing products or systems that were not fully tested, causing great difficulties for their customers.

Project management and personal time management are examples from business. Sometimes hurrying is the most reasonable way (write a report or letter from beginning to end without pauses). If, however, the work is big, it can be done only in parts (periodic action).

Principle 22. "Blessing in disguise." "Turn lemons into lemonade." *Use harmful factors (particularly, harmful effects of the environment or surroundings) to achieve a positive effect. Eliminate the primary harmful action by adding it to another harmful action to resolve the problem. Amplify a harmful factor to such a degree that it is no longer harmful.*

In the organization, complaints and destructive critique are negative "charges." They can be used to create positive change in the organization.

Virus attacks in computer networks are never good; but each time the system survives an attack, information is generated that makes the system better protected from the next attack. This is similar to the way the body works — surviving an illness generates antibodies that protect the victim from the next attack.

Principle 23. Feedback. *Introduce feedback (referring back, cross checking) to improve a process or action. If feedback is already used, change its magnitude or influence.*

The evolution of measurements and control is an example. On-line measurements and on-line control are increased. Quality control in production is improved by introducing the immediate measurement and control during the production process, compared to inspection after production. In business, systems for getting feedback from customers are being continuously improved.

Principle 24. Intermediary. *Use an intermediary carrier article or intermediary process. Merge one object temporarily with another (which can be easily removed).*

A neutral third party can be used as an intermediary during difficult negotiations. For sales promotion, an intermediary who is seen by the customer as an impartial expert can make recommendations.

Principle 25. Self-service. *Make an object or system serve itself by performing auxiliary helpful functions. Use resources including energy and materials, especially those that were originally wasted, to enhance the system.*

Some search engines use the frequency of use of a web site as the indicator of quality, so, the more often a site is used, the higher it rates on their recommendation list. This is a combination of feedback (Principle 23) and self-service.

Principle 26. Copying. *Instead of unavailable, expensive, or fragile objects, use simpler, inexpensive copies. Replace an object, system, or process with optical copies. If visible optical copies are already used, move to infrared or ultraviolet copies.*

Use a simulation instead of the object. This includes both physical and virtual simulations, such as virtual prototypes instead of physical ones and videoconferencing instead of travel.

Principle 27. Cheap disposables. Replace an expensive object with multiple inexpensive objects, compromising certain qualities (such as service life or appearance).

Disposable paper and plastic tableware; disposable surgical instruments; disposable protective clothing.

Controversial example: temporary employees or contractors instead of full-time employees.

Principle 28. Mechanical interaction substitution. Replace a mechanical means with a sensory (optical, acoustic, taste, or smell) means. Use electric, magnetic, and electromagnetic fields to interact with the object. Change from static to movable fields to those having structure. Use fields in conjunction with field-activated (e.g., ferromagnetic) particles.

The JIT manufacturing systems use kanban cards or objects such as portable bins to indicate visibly when supplies are needed.

In communication and business we also see clearly the increase of new interactions. Since the beginning of human society, communication has evolved from face-to-face communication to include writing, telegraph, telephone, fax, e-mail, videoconferencing, and other means.

Principle 29. Pneumatics and hydraulics. Use gas and liquid as parts of an object or system instead of solid parts (e.g., inflatable, filled with liquid, air cushion, hydrostatic, hydroreactive).

Business systems use this principle by analogy: "Water logic" vs. "rock logic" — fluid, flowing, gradually building up logic vs. permanent, hard-edged, rock-like alternatives.

Principle 30. Flexible shells and thin films. Use flexible shells and thin films instead of three-dimensional structures. Isolate an object or system from the external environment using flexible shells and thin films.

The thinnest film is a single molecule thick. Likewise, the thinnest organization structure is one employee thick. Get faster customer service by having the single employee customer service agent have all the necessary data easily available, so the customer deals only with the single, flexible "shell" of the organization instead of the whole bulky volume.

Heavy glass bottles for drinks are often replaced by cans made from thin metal (aluminum).

Principle 31. Porous materials. Make an object porous or add porous elements (inserts, coatings, etc.). If an object is already porous, use the pores to introduce a useful substance or function.

Think of the customer-facing layers of a company as a porous membrane that filters information flow both into and out of the organization.

Principle 32. Optical property changes. Change the color of an object or its external environment. Change the transparency of an object or its external environment.

Transparency is both a physical and a business term. Changing the transparency — increasing or decreasing it — is an important and often cheap way to improve business.

Principles 31 and 32 are frequently used together. Porous materials are semitransparent. A business analogy is the firewall in a computer system. The wall should be transparent for customers and other friends, while being impermeable for people who try to steal essential information.

Principle 33. Homogeneity. Make objects that interact out of the same material (or material with identical properties).

Business analogy: People may be more ready to buy things that remind them of familiar products than those that look very different. Movie sequels or series are a good example.

Principle 34. Discarding and recovering. Make portions of an object that have fulfilled their functions go away (discard by dissolving, evaporating, etc.) or modify them directly during operation. Conversely, restore consumable parts of an object directly in operation.

In business, the project organization is a good example of discarding and recovering. A good project should have an end. The organization will then be dissolved. The members can use their skills again in new projects. In all work, knowledge and skills are updated and improved directly during work and by retraining.

Principle 35. Parameter changes. Change an object's physical state (e.g., to a gas, liquid, or solid). Change the concentration or consistency. Change the degree of flexibility. Change the temperature.

In business situations, a parameter change is frequently realized as a policy change. In the past decade, many companies have increased the flexibility of employee benefit programs — instead of having one standard program, employees can design a mix of medical insurance, life insurance, pension plans, etc. Likewise, mass customization systems let customers have much more flexibility in designing products that exactly fit their needs.

Principle 36. Phase transitions. Use phenomena that occur during phase transitions (e.g., volume changes, loss or absorption of heat, etc.). The most common kinds of phase transitions are solid–liquid–gas–plasma, paramagnetic–ferromagnetic, and normal conductor–superconductor.

When businesses make structural changes (mergers, acquisitions, or internal changes), the accompanying phenomena are analogous to heat in a phase change — there is a lot of confusion! Constructive ways to use this period of disruption include finding new ways to align business systems with new strategies, forming new alliances with customers or suppliers, and getting rid of obsolete practices.

Principle 37. Thermal expansion. *Use thermal expansion (or contraction) of materials. Use multiple materials with different coefficients of thermal expansion.*

If employees are excited ("hot"), each can do more in the space that expands to exist between them.

Expand or contract marketing efforts depending on the product's "hotness" — rate of sales and profitability.

Principle 38. Strong oxidants. *Replace common air with oxygen-enriched air. Replace enriched air with pure oxygen. Expose air or oxygen to ionizing radiation. Use ionized oxygen. Replace ionized oxygen with ozone.*

Oxygen is used for bleaching pulp (for paper production). There are ideas and experiments to use ozone for bleaching.

Principle 39. Inert atmosphere. *Replace a normal environment with an inert one. Add neutral parts or inert additives to an object or system.*

Social analogy: indifference and neutrality. Ignore or neutralize negative and destructive actions (if you cannot turn them positive; see Principle 22). Use neutral arbitrators.

Principle 40. Composite materials. *Change from uniform to composite (multiple) materials and systems.*

The use of *nothing* (air or vacuum) as one of the elements of a composite is typical of TRIZ — nothing is an available resource in all situations! Examples include honeycomb materials (egg crates, aircraft structures), hollow systems (golf clubs, bones), and sponge materials (packaging materials, scuba diving suits), which combine this principle with Principle 31.

In business, we can speak of composite structures as well. Multidisciplinary project teams are often more effective than groups representing experts from one field. Multimedia presentations often do better in marketing, teaching, learning, and entertainment than single-medium performances. There are less tangible but not at all less important examples. Fanatic commitment to cleanliness is one famous feature of McDonald's. Consistent preparation of food is another major commitment. These are two principles, values, or fibers that tie together a loose organization.

The principle of composite materials or, more generally, composite systems, is a good conclusion for this section on using inventive principles. If you have a system, you can improve the result by combining it with another system. Innovative principles are also systems. Composite principles often do better than single ones.

Index

U

U charts, 114, 184
Univariate statistical tests, 135, 136
Updating
    final report and documentation contents, 185
    project charter, 65
Upper control limits (UCLs), 182

V

Value-added process analysis, 133; *see also*
            Analyze phase
Value-added vs. non value-added activities, 132
Value stream data and maps, 205, 209, 210, 215
Variable data, 77
    between and within samples, 144
    and sample size, 90
Variance, 82, 83; *see also* ANOVA
Vertical functions and horizontal processes, 40–42
Visual control, 6, 203
Voice of customer (VOC), 43–50
    critical-to-quality (CTQ) requirements, 48–50
        business, 50
        defined, 49–50
        integration into reporting systems, 50
    data collection, 45–48
        acting as customer for the day, 48
        complaint processes, 48
        direct observation, 47–48
        interviews and focus groups, 47
        surveys, 45–47
    define phase, 66
    identification of customer, 44–45
    importance, 43–44
    quality function deployment, 189

W

Waste
    lean production, 200, 201
    seven forms of, 195
    sources of, 5
Welch, Jack, 1, 7, 21, 22, 25
Western Electric, 6
Within-sample variation, 144
Work as process, 39–42
Workflow, process redesign, 159–160
Working population, 87
Worst case scenarios, 160

X

X and mR charts, 183
X-bar and R charts, 113–114, 183
X-bar and S charts, 114, 183
Xerox, 95

Z

Z and mR charts, 114
Zero Defects, 6, 17, 18
Zero Quality Control, 18
Z scores, converting test scores to, 83–84, 85
Z test, 137, 141, 146
    hypothesis testing, 136
    selection of appropriate test, 139